3-D AUDIO USING LOUDSPEAKERS

THE KLUWER INTERNATIONAL SERIES
IN ENGINEERING AND COMPUTER SCIENCE

3-D AUDIO USING LOUDSPEAKERS

by

William G. Gardner
Wave Arts, Inc.

KLUWER ACADEMIC PUBLISHERS
Boston / Dordrecht / London

Distributors for North, Central and South America:
Kluwer Academic Publishers
101 Philip Drive
Assinippi Park
Norwell, Massachusetts 02061 USA

Distributors for all other countries:
Kluwer Academic Publishers Group
Distribution Centre
Post Office Box 322
3300 AH Dordrecht, THE NETHERLANDS

Library of Congress Cataloging-in-Publication Data

A C.I.P. Catalogue record for this book is available
from the Library of Congress.

Copyright © 1998 William G. Gardner, portions copyright © 1997 Massachusetts Institute of Technology

All rights reserved. No part of this publication may be reproduced, stored in a retrieval system or transmitted in any form or by any means, mechanical, photocopying, recording, or otherwise, without the prior written permission of the publisher, Kluwer Academic Publishers, 101 Philip Drive, Assinippi Park, Norwell, Massachusetts 02061

Printed on acid-free paper.

Printed in the United States of America

CONTENTS

PREFACE vii

ACKNOWLEDGMENTS ix

1 INTRODUCTION 1
   1.1 Motivation: spatial auditory displays 1
   1.2 Ideas to be investigated 3
   1.3 Applications 4
   1.4 Organization of this book 6

2 BACKGROUND 7
   2.1 Sound localization in spatial auditory displays 7
   2.2 Loudspeaker displays 12

3 THEORY AND IMPLEMENTATION 21
   3.1 Introduction 21
   3.2 Head-related transfer functions 24
   3.3 Theory of crosstalk cancellation 44
   3.4 Bandlimited implementations 59

4 PHYSICAL VALIDATION 79
   4.1 Acoustical simulation 80
   4.2 Acoustical measurements 89

5 PSYCHOPHYSICAL VALIDATION 99
   5.1 Headphone and Loudspeaker Localization Experiments 100
   5.2 Validation of head tracking 116

	5.3 Dynamic head motion	127
	5.4 Discussion	131
6	DISCUSSION	133
	6.1 Contributions of this work	133
	6.2 Challenges to general use	134
	6.3 Directions for future work	137
A	INVERTING FIR FILTERS	141
REFERENCES		145
INDEX		153

PREFACE

This book is derived from my doctoral dissertation[†], completed in September 1997 at the MIT Media Lab under the supervision of Professor Barry Vercoe. Some revisions have been made to conform to stylistic requirements and to correct errors I have noticed since the dissertation was completed.

The topic concerns 3-D audio systems implemented using a pair of conventional loudspeakers. A well known problem with these systems is the requirement that the listener be properly positioned for the 3-D illusion to function correctly. This book proposes to use the tracked position of the listener's head to optimize the acoustical presentation, and thus produce a much more realistic illusion over a larger listening area than existing loudspeaker 3-D audio systems. Head tracking can be accomplished by applying pattern recognition techniques to images obtained from a video camera. Thus, an immersive audio environment can be created without donning headphones or other equipment.

The general approach to a 3-D audio system is to reconstruct the acoustic pressures at the listener's ears that would result from the natural listening situation to be simulated. To accomplish this using loudspeakers requires that first, the ear signals corresponding to the target scene are synthesized by appropriately encoding directional cues, a process known as "binaural synthesis," and second, these signals are delivered to the listener by inverting the transmission paths that exist from the speakers to the listener, a process known as "crosstalk cancellation." Existing crosstalk cancellation systems only function at a fixed listening location; when the listener moves away from the equalization zone, the 3-D illusion is lost. Steering the equalization zone to the tracked listener preserves the 3-D illusion over a large listening volume, thus simulating a reconstructed soundfield, and also provides dynamic localization cues by maintaining stationary external sound sources during head motion.

[†]Gardner, W. G. (1997). *3-D Audio Using Loudspeakers*, Ph.D. Thesis, MIT Media Lab.

This book discusses the theory, implementation, and testing of a head-tracked loudspeaker 3-D audio system. Crosstalk cancellers that can be steered to the location of a tracked listener are described. The objective performance of these systems has been evaluated using simulations and acoustical measurements made at the ears of human subjects. Many sound localization experiments were also conducted; the results show that head-tracking both significantly improves localization when the listener is displaced from the ideal listening location, and also enables dynamic localization cues.

This book should be of interest to researchers studying virtual acoustic displays, and to engineers developing the same. Much of the theory and experimental results presented are applicable to loudspeaker 3-D audio systems in general, not just head-tracked ones. It remains to be seen whether the head-tracked displays presented herein will find commercial applications; they certainly work in the laboratory.

Bill Gardner
Arlington, MA

ACKNOWLEDGMENTS

First and foremost, I would like to thank my advisor, Barry Vercoe, for providing constant encouragement and support during my graduate tenure at MIT. Barry's vision has resulted in the creation of the Machine Listening Group (formerly the Music and Cognition Group) at the MIT Media Lab, where researchers such as myself can freely pursue topics in the understanding and synthesis of music and audio. Barry has always encouraged me to "look at the big picture," a task which I have not mastered.

I can't imagine having a better doctoral committee than Bill Rabinowitz, David Griesinger, and Jean-Marc Jot. Each has a particular perspective that complements the others. Bill has been cheerful and supportive throughout; he has been particularly helpful with the psychoacoustic validation portion of this work. Bill also arranged for the use of the KEMAR and MIT's anechoic chamber, and he assisted me in building miniature microphones for making ear recordings. David has worked closely with me on previous projects, most notably a study of reverberation perception. We've met numerous times to discuss room reverberation, loudspeaker audio systems, and spatial hearing. He strives to find simple solutions to difficult problems; I hope I have emulated that ideal here. Jean-Marc's work in spatial audio synthesis, strongly grounded in theory, has been inspirational. Some of the key ideas in this work are the result of many illuminating email discussions with him.

My parents are no doubt responsible for my interest in science and for my choice of schooling at MIT. At a young age, they took me to MIT to see an exhibit of moon dust recently brought back from the lunar surface. From that time on I was certain that I would attend MIT; after a total of twelve years (both undergraduate and graduate) it's hard to believe that I'm finally out.

I would like to thank my colleagues in the Machine Listening Group for providing a stimulating and fun place to work. Thanks are extended to current group members Keith Martin, Michael Casey, Eric Scheirer, Paris Smaragdis, and Jonathan Feldman. Keith Martin, my current officemate, was closely involved with the early stages of this work, assisting me with the measurement of the KEMAR HRTF data. Keith is also to be thanked for providing a large music library for my late night listening pleasure. Eric Scheirer provided valuable proofreading assistance. Mike Casey, by using my systems in a number of audio productions, has both advertised my work and reminded me of the need for production-friendly designs.

Thanks are also extending to former members of our group, including Dan Ellis, Nicolas Saint-Arnaud, Jeff Bilmes, Tom Maglione, and Mary Ann Norris. Dan Ellis, my former officemate, deserves particular accolades. In addition to being a dear friend and providing emotional support during the initial stages of this work, he helped me solve countless problems ranging from computer glitches to theoretical issues. Dan is also credited with suggesting that I use the Tcl/Tk Toolkit to build a graphical interface for the sound localization experiment software; this was a huge time saver.

Connie Van Rheenen, Betty Lou McClanahan, Greg Tucker, and Molly Bancroft provided essential support at the Media Lab. Thanks are also extended to Bob Chidlaw at Kurzweil Music Systems, who first introduced me to the wonderful world of digital signal processing.

Finally, none of this would be possible without my wife Felice, who has been an unfaltering source of confidence, support, and love. This book is dedicated to her.

1 INTRODUCTION

In recent years there has been significant interest in the synthesis of immersive virtual environments. Applications for this technology include entertainment, communication, remote control, and simulation. It is essential that these simulations include a realistic recreation of the intended auditory scene. Four conditions necessary to achieve realism in a sound reproducing system have been described by Olson (1972):

- The frequency range must include all audible components of the sounds to be reproduced.
- The volume range must permit noiseless and distortionless reproduction of the entire range of intensities associated with the sounds.
- The spatial sound pattern of the original sound should be preserved in the reproduced sound.
- The reverberation characteristics of the original sound should be approximated in the reproduced sound.

Modern sound reproducing equipment can easily achieve the first two conditions. It is the latter two conditions that pose a challenge to the design of audio systems, particularly systems that are intended for extremely realistic spatial reproduction of audio.

1.1 MOTIVATION: SPATIAL AUDITORY DISPLAYS

Many different sound system technologies are in current use. These vary in the number and placement of transducers, which determine their spatial reproducing capabilities. The simplest, a monophonic sound system, is incapable of reproducing the spatial characteristics of sounds. All sounds, including reverberation, are reproduced from the location of the loudspeaker, resulting in an unnatural impression. Two-channel stereo sound systems are far superior, enabling the reproduction of sound images that are spatially distributed between the two loudspeakers. Stereo systems are generally incapable of rendering sound images originating from the sides or from behind the listener. The capabilities of stereo systems can be augmented by adding additional speakers to the sides or rear of the listener; the resulting *surround* systems are generally able to reproduce sound images anywhere in the horizontal plane surrounding the listener.

A *spatial auditory display* (also called a *virtual acoustic display* or a *3-D audio system*) is a system capable of rendering sound images positioned arbitrarily around a listener. There are two general approaches to building these systems. The first is to completely surround the listener with a large number of transducers, which enables the acoustic soundfield of the target scene to be exactly, or approximately, reproduced. The second is to reproduce only at the ears of the listener the acoustic signals that would occur in the natural listening situation to be simulated. This method, called *binaural audio,* is applicable to both headphone and loudspeaker reproduction. In theory, the binaural approach requires fewer transducers than the soundfield reproduction approach, because only the pressures at the two ears of the listener need to be reproduced. However, the binaural approach is not easily applied to loudspeaker systems intended for multiple listeners; the soundfield approach is better suited to this situation.

It is well known that the directional cues for sound are embodied in the transformation of sound pressure from the free field to the ears of a listener (Blauert, 1983). A head related transfer function (HRTF) is a measurement of this transformation for a specific sound location relative to the head, and describes the diffraction of sound by the torso, head, and external ear (pinna). A synthetic binaural signal is created by convolving a sound with the appropriate pair of HRTFs, a procedure called *binaural synthesis.* In order to correctly deliver the binaural signal to a listener using transducers, e.g. headphones, the signals must be equalized to compensate for the transmission paths from the transducers to the eardrums. This is accomplished by filtering the signals with the inverse of the transfer function that describes the transmission path, a procedure called *transmission path inversion.*

Headphones are often used for binaural audio because they have excellent channel separation, they can isolate the listener from external sounds and room reverberation, and the transmission paths from the transducers to the ears are easily inverted. However, when the synthesis of binaural directional cues is not tailored to the listener, headphone reproduction often suffers from in-head localization or front-back reversals, particularly for frontal targets (Begault, 1990; Wenzel, 1992). Headphones are also cumbersome and inconvenient.

An alternative to headphones is to use conventional stereo loudspeakers placed in front of the listener. In this case the transmission path inversion is accomplished by inverting the 2x2 matrix of transfer functions between the speakers and the ears. This is called *crosstalk cancellation* because it involves the acoustical cancellation of unwanted crosstalk from each speaker to the opposite ear. Binaural audio processed with a crosstalk canceller for loudspeaker playback, called *loudspeaker binaural*

3-D Audio Using Loudspeakers

audio, can sound quite realistic. In practice, loudspeaker binaural audio suffers less from in-head localization and poor frontal imaging than does headphone binaural audio. Moreover, the use of loudspeakers frees the listener from the requirement of donning headphones, and possibly being tethered by a wire. Unfortunately, loudspeaker binaural displays have a serious shortcoming: the crosstalk cancellation only functions in a fixed listener location, called the *sweet spot* or *equalization zone*. When the listener moves away from the equalization zone, the 3-D spatial illusion is lost. Furthermore, the spatial extent of the equalization zone is quite small; head translations as small as 10 cm or rotations of 10 degrees can noticeably degrade the spatial reproduction.

1.2 IDEAS TO BE INVESTIGATED

The central idea (strictly speaking, the "thesis") of this book is that the equalization zone of a crosstalk cancellation system can be steered to the position of a tracked listener, and that doing so greatly improves the simulation of a spatial auditory scene. Steering the equalization zone to the tracked listener should preserve the 3-D illusion over a large listening volume, thus simulating a reconstructed soundfield, and should also provide dynamic localization cues by maintaining stationary external sound images during head motion.

In this book, we will describe the theory and implementation of steerable crosstalk cancellers. We will then describe the objective performance of these systems as determined by simulations and acoustical measurements made at the ears of human subjects. We will also present the results of extensive sound localization experiments; these show that steering the equalization zone to the tracked listener both significantly improves localization when the listener is displaced from the ideal listening location, and also enables dynamic localization cues. The culmination of this research is a working implementation of a head-tracked 3-D loudspeaker audio system.

The approach we have taken is to implement binaural synthesizers and crosstalk cancellers based on a non-individualized head model. This model was obtained by making extensive HRTF measurements of a KEMAR (Knowles Electronic Mannequin for Acoustic Research[†]). Because the high-frequency features of HRTFs differ significantly across subjects, crosstalk cancellation is bandlimited to frequencies below 6 kHz. Above this frequency, the power levels of the binaural signals are adjusted in an effort to deliver the proper high-frequency powers to each ear. This power adjustment is obtained by expressing the crosstalk cancellation process in terms of power transfer. Therefore, we will present a hybrid approach to non-individualized crosstalk cancellation; at low frequencies the phases and magnitudes of the binaural signals are adjusted, at high frequencies only the power levels (magnitudes) are adjusted.

[†]Knowles Electronics, 1151 Maplewood Drive, Itasca, IL 60143.

Existing loudspeaker 3-D audio systems typically assume a centered listener and implement symmetric crosstalk cancellers. In general, a steerable crosstalk canceller must be asymmetric to deal with rotations of the listener's head. We will present a number of filter topologies that can implement both symmetric and asymmetric bandlimited crosstalk cancellers. The most computationally efficient crosstalk cancellers are based on recursive topologies that contain a parameterized model of head diffraction. By using a simplified head model, extremely low cost implementations are possible.

The objective performance of crosstalk cancellers can be described in terms of channel separation at the ears of a listener. We will present results of simulations that compare the performance of various crosstalk cancellers. The simulations also establish the spatial extent of the equalization zone, and verify that the equaliation zone can be steered using the described methods. We will also present the results of acoustical measurements made using both KEMAR and also miniature microphones inserted into the ear canals of human subjects. The results verify that the bandlimited crosstalk cancellers are on average effective at cancelling crosstalk up to 6 kHz.

Finally, we will present the results of an extensive set of sound localization experiments. The first set of experiments compare the performance of headphone and loudspeaker systems based on the KEMAR HRTFs; the results show that the loudspeaker system images frontal locations better than headphones. Additional experiments conducted with the loudspeaker system compare localization under tracked and untracked conditions when the listener is displaced from the ideal listening location; the results show that head tracking significantly improves localization performance, particularly when the listener is laterally translated or rotated. Another experiment tests localization using the loudspeaker system when the listener's head is rotating; the results show that head tracking enables dynamic localization cues.

Solutions to the problem of head tracking, i.e. determining the position of the listener's head, are not a goal of this work and will not be discussed in detail. Head tracking is a requirement of all binaural displays; the positions of sources must be rendered relative to an externally fixed reference, otherwise the auditory scene moves with the head rather than remaining stationary. With headphone displays, tracking is usually accomplished using an electromagnetic device affixed to the headphones (see Meyer et al., 1993, for a survey of position trackers). Forcing the listener to wear a tracking device conflicts with our goal of an unencumbered and untethered interface. We prefer techniques that use remote sensing, such as video based head tracking, which is an active area of research (Basu et al., 1996; Oliver et al., 1997).

1.3 APPLICATIONS

Head-tracked loudspeaker 3-D audio may be used in many applications that require or benefit from a spatial auditory display. Possible applications include interactive entertainment (e.g., video games, multimedia network applications), broadcast entertainment (e.g., television), communications (e.g., telephony, teleconferencing, flight control), simulation, remote control, and immersive interfaces. Applications for spatial auditory displays, and the need for general spatial control, have been detailed elsewhere (Wenzel, 1992; Begault, 1994; Shinn-Cunningham et al., 1997).

The concept of head-tracked loudspeaker 3-D audio is readily applied to the desktop computer interface. Computer monitors are routinely equipped with side mounted stereo loudspeakers; there is great interest in developing loudspeaker 3-D audio for desktop computers to enhance the immersive experience of video games and other multimedia applications without forcing the user to don headphones. Many implementations will likely use symmetrical crosstalk cancellers with a fixed equalization zone. However, these systems will not work properly when the user is incorrectly positioned. The use of a head tracked system has two advantages: not only will the system function when the user is incorrectly positioned, but user motions will actually enhance the listening experience. Head tracking can be accomplished using a video camera or other sensor mounted to the monitor. The video approach is attractive because applications such as teleconferencing already require a camera to be present. Both the head tracking and the audio processing (including binaural synthesis and crosstalk cancellation) can be implemented on the computer.

There are additional reasons why the desktop computer paradigm is ideally suited to our approach. First, desktop computers typically only have a single user. Theoretically, crosstalk cancellation systems can be built to service multiple simultaneous listeners (Bauck and Cooper, 1996), but it is enormously more complicated than the single listener situation. Second, a computer user tends to sit close to the computer monitor, and will therefore be close to the speakers. Consequently, the strength of any room relections will be relatively small compared to the direct sound coming from the speakers. Room reflections degrade the performance of loudspeaker 3-D audio systems, as we will later discuss. Finally, a computer user tends to orient their head towards the screen, which simplifies both the head tracking and audio reproduction tasks. We will later show that crosstalk cancellation works best when the loudspeakers are on opposite sides of the listener's medial plane; this constraint is conveniently accomodated by a user of a desktop computer.

The desktop computer is a particulary compelling example, but there are many other similar situations where a single user is facing some sort of display or terminal, such as personal laptop computers, arcade games, public telephones, automated teller machines (ATMs), and other kiosk-style interfaces. These are all well suited to implementing head tracked loudspeaker 3-D audio.

The constraints of car audio audio systems are also well suited to this technology. Car audio systems often have only a single listener, the driver. Because the position of the driver is known a priori, head tracking is not necessary. However, the techniques we will develop for delivering binaural audio using asymmetric loudspeaker geometries are directly applicable to car audio systems.

Television and home theatre are also potential application areas. Can this technology work in a living room? The answer to this often-asked question is "maybe." Living rooms pose considerable challenges because there may be multiple listeners, they are free to roam significantly, and they are relatively distant from the loudspeakers. We believe that the general technique we will study, i.e., delivering binaural cues to a listener by inverting the transmission paths, is applicable to any listening situation. However, the specific implementations we will discuss are intended for a single listener who is relatively close to the loudspeakers.

1.4 ORGANIZATION OF THIS BOOK

This book has six chapters. After this introduction, Chapter 2 reviews related work in loudspeaker 3-D audio and discusses alternative technologies for implementing spatial auditory displays. Chapter 3 describes in detail the theory and design of crosstalk cancellers. This includes discussions of the KEMAR HRTF measurements, head diffraction models, and the high-frequency power model. Chapter 4 discusses the simulations and physical measurements used to validate the objective performance of the systems described in Chapter 3. Chapter 5 describes the sound localization experiments conducted to validate the subjective performance of the systems described in Chapter 3. Finally, Chapter 6 discusses the results of this study, and suggests areas of future work.

2 BACKGROUND

This chapter will review two important topics relevant to our study: sound localization in spatial auditory displays, and loudspeaker display technologies.

2.1 SOUND LOCALIZATION IN SPATIAL AUDITORY DISPLAYS

Sound localization by humans has been studied extensively (e.g., Mills, 1972; Durlach and Colburn, 1978; Blauert, 1983; Oldfield and Parker, 1984a, 1984b, 1986; Yost and Gourevitch, 1987; Wightman and Kistler, 1989a, 1989b; Makous and Middlebrooks, 1990; Middlebrooks and Green, 1991; Gilkey and Anderson, 1997). Recent interest in spatial auditory displays has prompted the study of sound localization specifically related to this technology (e.g., Wightman and Kistler, 1989a, 1989b; Wenzel, 1992; Wenzel et al., 1993, Moller et al., 1996a; Gilkey and Anderson, 1997). Here we will briefly summarize some of the important findings, especially those that are relevant to our study.

2.1.1 Interaural cues

It has long been known that the principal cues for sound localization, particularly localization to the left or right, are the time and level differences at the ears of the listener (Rayleigh, 1907). Rayleigh's "duplex theory" states that low frequencies are localized using time (phase) cues, and high frequencies are localized using interaural level (intensity) cues. An *interaural level difference* (ILD) will lateralize a sound towards the ear with the greater intensity; this cue works at all frequencies, but natural head shadowing does not attenuate low frequencies substantially unless the source is very close to the head. *Interaural time delay* (ITD) cues are effective at localizing low-frequency sounds; they principally operate at frequencies below about 1500 Hz. The question of which ear has leading phase can be unambiguously determined for frequencies below about 700 Hz. The ability of neurons in the auditory periphery to phase lock to amplitude modulations of high-frequency carriers enables a high-frequency ITD cue based on the time difference of the amplitude envelopes; however, this cue is rather weak compared to the lower frequency phase cue (Durlach and Colburn, 1978, pg. 399).

The relative salience of these cues can be tested in various ways. Mills (1972) has shown that the sensitivity of azimuth changes about the medial plane can be explained by the sensitivity to ITD for frequencies below 1500 Hz, and by sensitivity to ILD for frequencies from 1500 Hz to 6 kHz. ITD and ILD cues can be pitted against one another in cue trading experiments (e.g., Mills, 1972; Durlach and Colburn, 1978). The cue trading ratios vary considerably depending upon the region being tested, the stimuli, and experimental conditions. Another approach is to perform localization experiments using binaural stimuli synthesized with conflicting ITD and ILD cues, Using this approach,Wightman and Kistler (1992) have shown that low-frequency ITD cues dominate ILD cues; in other words, when the stimuli contain low frequencies, the position of the auditory image is determined by the ITD cue regardless of the ILD cues. This finding is particularly relevant to our study.

Both ITD and ILD cues are frequency dependent. When the ITD is defined in terms of the total phase difference at the ears (we discuss an alternative definition in Chapter 3), it is seen that the ITD at frequencies below 1500 Hz is about 3/2 larger than the ITD above this frequency, which agrees which spherical diffraction theory (Rayleigh, 1945; Kuhn, 1977, 1987). The frequency dependence of the ILD is rather complicated: at frequencies below about 6 kHz, the ILD of lateral sources decreases with increasing frequency due to the frequency dependence of head shadowing; at higher frequencies the spectral features are quite complicated due to the filtering effects of the external ear. Plots of ITDs and ILDs are given in Chapter 3.

Any head model that is axially symmetric along the interaural axis, such as a sphere, will lead to interaural cues that depend only on the distance of the source and the angle of the source with respect to the interaural axis. The distance dependence is only significant for sources close to the head. The interaural ambiguity leads to the so-called *cone of confusion* (Mills, 1972), which is the locus of equally possible source locations as determined by an interaural cue. Confusion errors are a well known sound localization phenomenon, consisting primarily of front-back confusions, which occur frequently with fixed head listening (Rayleigh, 1907; Oldfield and Parker, 1984a; Makous and Middlebrooks, 1990; Wightman and Kistler, 1989b). Front-back confusions are often accompanied by an increase in elevation localization error (Makous and Middlebrooks, 1990). Up-down confusions have also been reported (Wenzel et al., 1993), though these occur less frequently than front-back confusions. It is generally accepted that spectral cues are used for front-back and elevation localization; however, it is not well understood how the various interaural and spectral cues are integrated to form a localization judgement.

2.1.2 Spectral cues

The spectral modification of sound due to interaction with external ear is a well known sound localization cue. Batteau (1967) proposed that directional cues are encoded by multipath reflections off the pinna that sum at the ear canal; the pattern of the delayed reflections depends on the direction of the sound source. Several studies have tested sound localization when various anatomical features of the external ear are occluded (Gardner and Gardner, 1973; Oldfield and Parker, 1984b). These studies show that pinna cues contribute significantly to both elevation localization (for both medial and non-medial sources) and also front-back discrimination. Other studies have tested medial localization by presenting stimuli that have been filtered in various ways to alter the spectral content (Blauert, 1969/70; Gardner, 1973; Hebrank and Wright, 1974b; Asano et al., 1990). Blauert (1969/70) determined that the localization of 1/3-octave noise stimuli on the median plane is primarily determined by the stimulus frequency, independent of the location of the source. Furthermore, the HRTFs for locations chosen in response to certain frequencies were seen to have spectral maxima ("boosted bands") at those same frequencies. This result suggests that spectral maxima in HRTFs determine the location of sounds on the medial plane, and that certain features are similar across subjects. The importance of spectral peaks as localization cues has been shown by other authors (e.g., Butler, 1997).

Hebrank and Wright (1974b) tested medial localization using lowpass, highpass, bandpass, and bandcut filtered noises. The results showed that sound spectra from 4 to 16 kHz were necessary for medial localization. A number of spectral cues for "front," "above," and "behind" were identified, including peaks and notches at various frequencies. The authors hypothesized that the principal frontal elevation cue is a notch created by interference with a reflection off the posterior wall of the concha; the notch frequency ranges from approximately 6 to 13 kHz as elevation increases from -30 to +30 degrees. The importance of this notch for elevation localization has also been stressed by other authors (Bloom, 1977; Butler and Belendiuk, 1977; Lopez-Poveda and Meddis, 1996).

Notch and peak spectral features are not independent; the presence of a notch is the absence of a peak and vice-versa. Both features can be incorporated into a general spectral shape cue (Middlebrooks, 1992, 1997). Although spectral shape cues are principally used for front-back discrimination and elevation localization, they are also potentially useful for horizontal localization. Good evidence for this is that monaural listeners can localize horizontal localizations fairly well, although they have access to only a monaural spectral shape cue (Middlebrooks, 1997).

Because human heads are not axially symmetric along the interaural axis, interaural differences, particularly the frequency dependent ILD, change systematically as a function of position along cones of confusion (Duda, 1997). Frequency dependent ILDs, also called *interaural difference spectra*, are attractive as a possible localization cue because they don't depend on the source spectrum. Searle et al. (1975) proposed interaural difference spectra caused by pinna disparities as a medial localization cue. This idea was disputed by Hebrank and Wright (1974a), who demonstrated that subjects could localize medial locations just as well with monaural listening conditions (one ear plugged) as with binaural listening conditions, provided the subjects received sufficient training. The relative importance of binaural versus monaural spectral cues is not well understood.

2.1.3 Individualized HRTFs

Several studies have been made of sound localization using synthetic binaural stimuli presented over headphones. Wightman and Kistler (1989a, 1989b) demonstrated that localization performance over headphones is essentially the same as with free-field conditions when the stimuli are synthesized using the subject's own HRTFs, i.e., using *individualized* HRTFs. Wenzel et al. (1993) studied localization over headphones using non-individualized HRTFs taken from the ears of a good human localizer. The use of non-individualized HRTFs increased elevation localization errors and front-back reversals, but horizontal localization was largely unaffected. Front-to-back reversals were more frequent than back-to-front reversals. These results are in general agreement with a recent study by Moller et al. (1996a). The increased errors using non-individualized HRTFs are explained by the fact that elevation localization and front-back resolution depend on high-frequency spectral cues which vary considerably across subjects.

2.1.4 Externalization

The subject of in-head localization is also important to the study of spatial auditory displays. It has been shown that both headphones and loudspeakers are capable of generating sounds that are perceived either in the head or outside the head (Toole, 1970; Plenge, 1974). Thus, the externalization of auditory images is independent of the type of transducers used, although in-head localization is more frequently encountered with headphone listening. The externalization of auditory images is substantially dependent on the following cues:

1. Individualized pinna cues (Hartmann and Wittenberg, 1996),
2. Reverberation cues (Begault, 1992),
3. Dynamic localization cues (Wallach, 1939, 1940; Wenzel, 1995; Wightman and Kistler, 1997),
4. Corresponding visual cues (Durlach et al., 1992).

It is unclear whether individualized pinna cues are necessary for external localization. Hartmann and Wittenberg (1996) studied how various spectral manipulations of sound can degrade the perception of externalized images. An important finding of this study is that the correct reproduction of interaural difference spectra is insufficient to maintain externalized sound images; rather, the monophonic spectral cues at each ear must be reproduced correctly, which would support the notion that individualized pinna cues are needed. Many studies have shown that non-individualized HRTFs can synthesize external images (Plenge, 1974; Begault, 1992; Moller et al., 1996a); however, in all these cases the stimuli that were externally perceived contained real or artificial room reverberation. The conclusion is that non-individualized HRTFs can be used to generate externally perceived images when other externalizing cues are present.

2.1.5 Dynamic localization

It is well known that head movement is important for sound localization. Wallach's (1939, 1940) conducted a series of localization experiments using a clever apparatus that allowed the manipulation of dynamic localization cues in response to head motion. Head rotation was coupled to a rotary switch that connected the sound source to one of a set of loudspeakers placed in an arc in front of the listener. The relationship betwen interaural cues and head position could thus be controlled by appropriately mapping head rotations to loudspeakers. For instance, the sound could be made to eminate from the loudspeaker directly in front of the listener for all head roations. The only stationary sound source that would yield this relationship between head rotation and interaural changes is an overhead sound. Listeners would initially report the sound image as frontal, but subsequent head rotation would cause the image to jump to overhead and remain there. It was also possible to synthesize rear sources by suitable mapping of rotations to interaural changes. As with the overhead case, listeners would initially localize the image as frontal, but head motion would cause the image to jump to the rear. These experiments convincingly demonstrated that dynamic localization cues dominate pinna cues for front-back and elevation localization.

Thurlow and Runge (1967) studied localization during induced head rotation; the results demonstrated that the induced head rotation was especially effective in reducing horizontal localization error and somewhat effective at reducing vertical localization error. Thurlow et al. (1967) studied subjects' head motions during sound localization. Subjects were instructed to keep their torsos still but were permitted to move their heads freely to aid in localization. Head rotation about the vertical axis was the most frequently observed motion, often towards the location of the sound, and the largest average rotations (42 degrees) occurred when localizing low-frequency sounds. The study clearly shows that individuals will freely move their heads to localize sound better. It should be noted that the observed rotations are not only larger than the maximum allowable rotation (10 degrees) reported for crosstalk cancelled loudspeaker audio, but are also larger than typical stereo speaker angles (30 degrees).

Studies of dynamic sound localization have also been made with synthetic binaural stimuli delivered over headphones. Adding dynamic head-tracking to a headphone display greatly decreases front-back reversals (Boerger et al., 1977; Wightman and Kistler, 1997). The study by Wenzel (1995) shows that changing ILD cues in response to head motion is more important for localization accuracy than changing ITD cues. This finding is quite relevant to the design of a head tracked loudspeaker display.

2.2 LOUDSPEAKER DISPLAYS

In this section we review loudspeaker display technologies with emphasis on their relation to our study.

2.2.1 Stereo

Strictly speaking, stereo refers not to the use of two channels, but the ability of the sound system to reproduce three-dimensional sound. However, we will use the term stereo to denote two-channel reproduction. Stereo systems have been in use for decades, and have been extensively studied (e.g., AES, 1986). Essentially, the stereo technique relies on the ability to position a sound *between* the two loudspeakers by adjusting the amplitude and/or delay of the sound at each speaker. Individually, these techniques are called *intensity panning* and *time panning*, respectively.

Time-panned stereo is problematic. Using equal amplitude broadband signals, delaying one channel by less than a millisecond is sufficient to move the auditory event to the opposite speaker (Blauert, 1983, pg. 206). This result is from de Boer's experiments using a conventional stereo arrangement (speakers at ± 30 degrees) with the listener's head immobilized. Similar experiments show that time panning depends greatly on the signal, and is generally not effective with narrowband signals. Moreover, any attempt to use time panning is defeated by a mobile listener, because lateral motions of a few feet can create speaker to head delay differences which overwhelm the short delays intended for panning. If time-panning were to be used, it would require tracking the listener in order to adapt the panning delays correctly.

Intensity panning is far more effective and robust than time panning. About 25 dB of level difference is sufficient to move the auditory event completely to the stronger speaker (Blauert, 1983, pg. 206). Intensity panning works fairly consistently with different signal types, even with narrowband signals, although high-frequency sinusoids give degenerate results. Unlike time-panning, intensity panning is still effective when the listener is off-axis. The success of intensity panning has led to a number of coincident microphone techniques for stereo, which date back to the 1930's with Blumlein's pioneering work (Blumlein, 1933, 1958; Lipshitz, 1986; Griesinger, 1987; Heegaard, 1992).

Stereo techniques can be analyzed using phasor methods (Bauer, 1961a), which assume the signals are steady state sinusoids, and thus are completely specified by a

3-D Audio Using Loudspeakers

complex value. The signal at one ear is the sum of the same-side (ipsilateral) speaker phasor and the opposite-side (contralateral) speaker phasor, which is delayed (rotated) to account for the interaural time delay. Phasor analysis demonstrates that intensity differences between the two speakers result in ear signals which have the same intensity but different phase; conversely, speaker phase differences yield interaural intensity differences at the ears (Lipshitz, 1986). Many papers have been written on phasor analysis and appropriate head models (e.g., Cooper, 1987), but this analysis technique is only valid for low-frequency sinusoids and for a fixed, on-axis listener position, and is not applicable to a mobile listener or broadband signals.

Stereo techniques may also be explained as a consequence of *summing localization*, whereby a single auditory event is perceived in response to two sources radiating coherent signals (Blauert, 1983). When either one of the sources radiates a locatable signal, the auditory event appears at the location of the source. When both sources radiate the same signal in some amplitude proportion, a single auditory event is perceived at a location between the two sources, even though the actual ear signals are not entirely consistent with this perception. It is clear that the auditory system is unable to separately identify the two sources, and assigns a best guess location to the auditory event in the presence of conflicting cues.

2.2.2 Crosstalk cancellation

This section reviews previous studies of crosstalk cancellation systems. Some of the techniques introduced here will be described in greater detail in Chapter 3.

Crosstalk cancellation is a technique for sending arbitrary, independent signals to the two ears of a listener from conventional stereo loudspeakers; it involves canceling the crosstalk that transits the head from each speaker to the opposite ear. The technique was first introduced by Bauer (1961b), put into practice by Schroeder and Atal (1963), and later used by Schroeder to reproduce concert hall recordings for a comparative study (Schroeder 1970, 1973; Schroeder et al., 1974). Essentially, the transfer functions from the two speakers to the two ears form a 2x2 system transfer matrix. To send arbitrary binaural signals to the ears requires pre-filtering the signals with the inverse of this matrix before sending the signals to the speakers. The inverse filter, or the *crosstalk canceller*, as we will call it, is a two-input, two-output filter which Schroeder implemented using a lattice topology (described in Chapter 3). The filter functions were derived from head responses measured using a dummy head microphone. For the comparative study, binaural impulse responses of concert halls were convolved with anechoic music to create binaural signals. These were filtered with the crosstalk canceller and presented to a listener seated in an anechoic chamber with loudspeakers at ± 22.5 degrees. Schroeder described the result as "nothing less than amazing" (Schroeder, 1973). Listeners could perceive sound originating from all directions around them, although no localization experiments were done. Schroeder reported that immobilizing the head was not necessary, but that head rotations of ± 10 degrees were sufficient to ruin the spatial illusion.

The usual method of creating crosstalk cancelling filters is to invert head responses obtained by direct measurement or modeling. Damaske (1971) described an alternative method whereby the cancellation filters were specified through a calibration procedure. A subject was seated in front of stereo speakers, spaced at ±36 degrees, which emitted the same bandpass filtered noise. The subject was instructed to adjust the delay and gain (including inversion) of the right-hand speaker to move the noise so that its apparent direction was directly to the left of the subject. This procedure was repeated using 1/3-octave noises chosen at 10 center frequencies from 400 Hz to 10 kHz. The results specified a "90-degree" filter, which was superposed with the mirror 90-degree filter to build a symmetric crosstalk canceller. The filter was based on the results obtained from one subject. Damaske conducted many localization experiments under various playback conditions using binaural speech recorded from a dummy head microphone. Subjects were properly positioned but instructed to keep still. The results showed excellent localization for all azimuths in the horizontal plane. Vertical localization in the median plane was also good. In both cases, sources directly to the rear were occasionally perceived as being frontal. As the reverberation in the playback room increased, so too did the frequency of front-back reversals for all sources from rear azimuths, however frontal sources were always perceived correctly. Damaske also showed that moving subjects laterally off axis by as little as 10 cm caused the localization performance to degenerate to stereo performance, in which case sources could only be localized between the speakers.

Cooper and Bauck (1989) simplified the crosstalk canceller by exploiting the symmetry of the listening situation. This yields a crosstalk canceller implemented with only two elementary filters, one that operates on the sum of the left and right binaural inputs (L+R), and one that operates on the difference of the inputs (L-R). This topology, called a *shuffler*, has been used historically to process stereo recordings (Blumlein, 1933, 1958; Gerzon, 1994). In contrast to the four filters required for the lattice implementation of the general 2x2 case, the shuffler implementation only requires two filters that have a particularly simple form, which will be described in Chapter 3. In an effort to match the crosstalk canceller to an average listener, Cooper and Bauck based their filters on a spherical head model (Cooper, 1982).

Several other filter topologies for implementing symmetric crosstalk cancellers have been proposed. Iwahari and Mori (1978) described a recursive circuit whereby each output channel is fed back to the opposite channel's input via a filter consisting of the ratio of the contralateral to ipsilateral HRTFs. This recursive topology and related topologies will be discussed in Chapter 3. Sakamoto et al. (1981) described a circuit consisting of a "ratio" filter and a "common" filter. The ratio filter, which was applied to only one channel, affected crosstalk cancellation; the common filter, which was applied to both channels, affected the overall equalization. The filters were based on measurements of a dummy-head microphone. Experiments with this system demon-

3-D Audio Using Loudspeakers

strated good horizontal and vertical localization performance (Sakamoto et al., 1982). It was shown that disabling the common circuit greatly increased back-to-front reversals, indicating that the overall equalization is important for correctly perceiving rear sources.

Other authors have described crosstalk cancellers based on HRTFs measured from humans or dummy head microphones (Moller, 1989; Kotorynski, 1990; Jot, 1992; Koring and Schmitz, 1993). Kotorynski (1990) focused on the problem of designing minimum-phase crosstalk canceller filters based on non-minimum-phase HRTFs. A symmetric system, implemented with 100-tap FIR shuffler filters, yielded better than 20 dB of broadband channel separation. The crosstalk cancellers inverted the full audio range, and the filters therefore contained sharp peaks at high frequencies caused by pinna-related notches in the HRTFs. Kotorynski (1990) also implemented asymmetric crosstalk cancellers using a lattice filter topology. Jot (1992) created efficient shuffler filters based on a feedback loop containing a delay and a lowpass filter. The lowpass filter was derived from the ratio of the minimum phase parts of the ipsilateral and contralateral HRTFs, a method which will be described in Chapter 3.

Koring and Schmitz (1993) described crosstalk cancellers based on individualized HRTF measurements and implemented using a lattice filter topology. The authors noted exceptionally high fidelity reproduction of binaural recordings in the case of identity listening, i.e., when the listener was the same individual measured to create the crosstalk canceller and also used for binaural recording. However, non-identity listening revealed unnatural timbre artifacts, attributed to differences in the fine structure of the listeners' HRTFs. To prevent the equalization of small dips in the HRTFs, the frequency magnitudes of the HRTFs were first smoothed using a 1/3-octave averaging filter. This greatly lessened the timbre artifacts associated with non-identity listening without significantly degrading localization performance. A number of sound localization experiments were performed using identity listeners. In anechoic conditions, both horizontal and vertical localization performance was excellent. Localization performance degraded in increasingly reverberant conditions, exhibiting increased front-back reversals and elevation errors. A "typical" crosstalk canceller was also selected by testing the localization performance of a set of individualized crosstalk cancellers using a set of non-identity listeners and choosing the system with the best overall rating. However, localization results using the typical system were not shown.

Crosstalk cancellation is related to the problem of equalizing dummy head recordings for loudspeaker playback. A particular problem with dummy head recordings is that they have little low-frequency separation, and the subtle time differences at low frequencies are incapable of indicating direction when reproduced through loudspeakers. Griesinger (1989a) describes shuffler filters which apply a bass boost to the L-R difference signal, a technique first proposed by Blumlein (1933, 1958) for equalizing stereo recordings. This increases the spaciousness of the reproduction and allows low-frequency sounds to be localized. At low frequencies, crosstalk cancellation is equivalent to a phase-corrected L-R bass boost (Griesinger, 1989b). Increasing the low-frequency separation pumps energy into the low-frequency lateral modes of the room (Griesinger, 1986), and this lateral energy at low frequencies greatly contributes to the sensation of spaciousness (Morimoto and Maekawa, 1988).

The problem of transmission path inversion has been extended to multiple speakers and listeners (Yanagida et al., 1983; Bauck and Cooper, 1992, 1993, 1996; Abe et al., 1995). Matrix notation is used to specify the general problem of sending P program signals through M loudspeakers to L ears, where the desired transfer function from program signals to ears is given. The general solution involves a pseudoinverse of the LxM system transfer matrix **X**. When M > rank(**X**), the system is underdetermined, and an infinite set of solutions exist. A least squares solution minimizes total signal power delivered to the loudspeakers (Bauck and Cooper, 1992, 1993, 1996). In the underdetermined case, we note that the additional degrees of freedom could be used to minimize the spatial gradients of sound pressure at the equalization zone (Asano et al., 1996); doing so should increase the spatial extent of the equalization zone. When L > rank(**X**), the system is overdetermined, and no exact solutions exists. In this case, a least squares solution minimizes the squared error. Abe et al. (1995) proposed solving for additional points in the vicinity of the equalization zone in an effort to increase its spatial extent; the technique was effective for certain loudspeaker geometries.

A different approach to multichannel crosstalk cancellation, used by Iwahara and Mori (1978), is to partition the set of speakers into pairs; each speaker pair becomes a separate 2x2 crosstalk canceller. It is then possible to pan the source signal to a specific crosstalk canceller for optimum reproduction depending on the desired target location. For instance, a four-channel system could be composed of a front crosstalk canceller using the front-left and front-right speakers, and a rear crosstalk canceller using the rear-left and rear-right speakers. Frontal sounds are panned to the front system and rear sounds are panned to the rear system. This strategy would be expected to reduce front-back reversals when using non-individualized crosstalk cancellers.

2.2.3 Inverse filtering of room acoustics

The crosstalk cancellers described in the preceding section invert only the listener's head response, and do not compensate for the acoustics of the listening space. It is possible to invert a room's acoustic impulse response with a causal, stable filter only when the room response is minimum phase (Neely and Allen, 1979). However, room responses are seldom minimum phase, and therefore it is necessary to incorporate significant modeling delay into the inverse filter in order to obtain an approximate inverse (Mourjopoulos, 1985). This works at the exact point in the room where the impulse response measurement was taken, but all other points in the room are subject to the pre-response of the inverse filter (i.e., the response prior to the modeling delay), which sounds objectionable. For this reason, most techniques equalize only the minimum phase portion of the room response, leaving the excess phase portion unchanged (Craven and Gerzon, 1992).

A method for exactly inverting acoustic impulse responses in a room is described by Miyoshi and Kaneda (1988). Because of the non-minimum phase nature of room responses, it is not possible to realize an exact inverse when the number of sources is equal to the number of equalization points one wishes to control. However, by adding one extra source (and hence one extra transmission path to each equalization point) it becomes possible to realize an exact inverse using FIR filters. This principle is called the multiple-input/output inverse theorem (MINT). This important result is obtained when the transmission convolutions are formulated as a matrix multiplication, and then the inverse of this matrix yields the set of FIR inverse filters. The MINT principle follows from the requirement that the system transfer matrix be square in order to be invertible.

Elliot and Nelson have studied the design of adaptive filters for multichannel sound reproduction systems (Elliot and Nelson, 1985; Elliot et al., 1987; Elliot and Nelson, 1989; Nelson et al., 1992). A multiple error least mean square (ME-LMS) algorithm is developed to adaptively design a matrix of FIR inverse filters. These filters can perform crosstalk cancellation and also equalize for the room response. A practical implementation would require that the listener have microphones positioned in the ear canals in order to adapt the inverse filters to the optimal solution, although the microphones could be removed after the filters converge to a solution. Nelson et al. (1995) have since demonstrated that the MINT and the ME-LMS method are equivalent, and have derived conditions that must be fulfilled for an exact inverse to exist. The spatial extent of the resulting equalization zone was shown to depend on the acoustic wavelength of the highest frequency of interest.

2.2.4 Ambisonics

The Ambisonics B format is a four channel audio format that encodes spatial audio recorded using a soundfield microphone (Gerzon, 1985; Farrar, 1979). The four channels consist of the omnidirectional pressure at the recording position, plus the dipole (pressure gradient) responses along the x, y, and z axes. The playback strategy is frequency dependent (Gerzon, 1992): at frequencies below 700 Hz, the omnidirectional pressure and pressure gradients at a point in space are reconstructed; at higher frequencies, the total energy and the directional flow of energy are reproduced. The different strategies for low and high frequencies are accomodated by placing suitable shelving filters in the decoders (Gerzon, 1992).

Ambisonics B format is called a first-order system, because (at low frequencies) it reconstructs the zero-order pressure and first order pressure gradients at a point. Higher-order systems are theoretically possible and would seek to reconstruct higher-order gradients. Ambisonics has been described as decomposing the soundfield into spherical harmonics (Bamford and Vanderkooy, 1995), which form an orthogonal basis for representing spherical sound sources (Morse and Ingard, 1968). Jot has described the technique as reconstructing a soundfield around a point by polynomial approximation†, a technique closely related to the derivative constraints discussed by Asano et al. (1996).

In practice, Ambisonics systems suffer from a variety of problems. As with crosstalk cancellers, there is an optimum listening position, or sweet spot. All speakers reproduce the omnidirectional component, and this causes a distinct timbral artifact as the listener moves his head near the sweet spot, due to the reinforcement and cancellation of sound at a periodic set of frequencies. When the listener is positioned correctly, the omnidirectional component can be perceived as being located inside the head. The same problems are encountered when reproducing monophonic recordings over stereo loudspeakers. Another deficiency of the Ambisonics encoding is that sound originating from one direction will be reproduced by many loudspeakers, even if one loudspeaker corresponds exactly to the desired direction.

Despite these difficulties, Ambisonics is a compact and useful format for representing spatial audio. Spatial manipulations of the encoded audio, such as rotation about an axis, can be easily performed; this is particularly useful for adjusting sound playback in response to head rotation when using a headphone spatial auditory display. Ambisonics also permits manipulation of the sweet spot position to adjust for the listener's location (Burraston et al., 1997).

2.2.5 Discrete Surround

Discrete surround systems consist of many loudspeakers surrounding the listener. Early quadraphonic sound systems used four speakers in a square arrangement (Olson, 1972). A well known problem with quadraphonic systems is the difficulty in reproducing directions between the speakers, particularly lateral directions (Theile and Plenge, 1977). Six channel systems, with speakers spaced at 60 degrees, are capable of fairly accurate reproduction of all directions on the horizontal plane. An extremely popular new format for home theaters is the 3/2 format, with three frontal speakers and two rear surround speakers (Theile, 1993; Greisinger, 1996). This format allows for accurate frontal localization (important when the audio corresponds to sources on screen), and provides rear channels for reproducing room acoustics and other spatial sounds. These systems provide excellent directional reproduction to multiple listeners in a wide listening area. The main problem with these systems is the requirement of providing speakers to the side or behind the listener, which is not always practical.

Discrete surround systems use separate strategies for rendering directional sounds and diffuse reverberation (Gardner, 1992; Jot, 1996, 1997; Gardner, 1998). Directional

---

†Jean-Marc Jot, personal communication, 1996.

3-D Audio Using Loudspeakers

sounds are positioned using intensity panning between adjacent loudspeakers in the array (Pulkki, 1997). Diffuse reverberation is usually rendered by all speakers. Ideally, each speaker's reverberation is uncorrelated with the others, but it is sufficient to provide several channels of uncorrelated reverberation and assign these to the speakers by spatial interleaving (Griesinger, 1991).

Several authors have described large scale discrete systems that contain 50 or more speakers arranged in a sphere (Meyer et al., 1965; Kleiner, 1981), intended for extremely realistic simulation of concert halls. If the number of channels is further increased, it becomes possible to exactly reconstruct the original soundfield. This technique is called holographic audio (Berkhout, 1988) or wave field synthesis (Berkhout et al., 1993; Start et al., 1995). The theory is based on the Kirchhoff-Helmholtz integral which states that an arbitrary sound field within a closed volume can be generated with a distribution of monopole and dipole sources on the surface of the volume, provided the volume itself does not contain sources (Boone et al., 1995). Although no practical implementations of this technology exist, it represents the theoretical limiting case of exact soundfield reproduction.

3 THEORY AND IMPLEMENTATION

3.1 INTRODUCTION

This chapter describes in detail the methods we will use to implement head-tracked loudspeaker 3-D audio. After this introduction, we will describe the HRTF measurements we will use for head models, then will describe the mathematical theory of crosstalk cancellation. Finally, we will discuss our hybrid approach to crosstalk cancellation: the combination of a bandlimited crosstalk canceller with a high-frequency power compensation circuit.

3.1.1 Binaural audio using loudspeakers

The approach taken in this book is to deliver binaural audio to the ears of the listener, and therefore only the acoustic pressures at the eardrums of the listener are considered fundamentally important. This approach requires far fewer transducers than a system that attempts to reconstruct a complex soundfield within a volume of space. Existing loudspeaker systems that deliver binaural audio to a listener have the serious constraint that the listener may not move. Our approach is to track the listener and adjust the loudspeaker signals to maintain the binaural transmission, thus simulating a reconstructed soundfield in a volume of space.

The system is created by combining a binaural synthesizer with a circuit that inverts the acoustic transmission path to the ears. The primary goal of the transmission path inversion is to eliminate crosstalk from each speaker to the opposite ear, and these circuits are called crosstalk cancellers. Although it is possible to merge the operation of the binaural synthesizer and crosstalk canceller into a single filter operation, there are many reasons for logically separating these operations:

- The binaural synthesizer and crosstalk canceller both require head models. It is possible to use different head models for each function, and each may be individualized or non-individualized.

- The head model used by the binaural synthesizer is intended to be perceptually correct, whereas the head model used by the crosstalk canceller must be acoustically correct.
- Efficient implementations can result from suitable factorizations of the separate systems.
- Reverberation is properly handled by bypassing the binaural synthesizer and using only the crosstalk canceller.
- Headphone compatibility is easily achieved when the systems are separately implemented; the headphones can be driven from the output of the binaural synthesizer.
- When analyzing the performance of the total system, errors may be attributed to deficiencies in either the binaural synthesizer or crosstalk canceller.

It is therefore desirable to separately implement the binaural synthesizer and crosstalk canceller. Our implementations will use head models based on measurements of a KEMAR dummy head microphone for both the binaural synthesis and the transmission path inversion.

3.1.2 Head tracking

Head tracking is necessary if the listener is not to be constrained to a single fixed location. Both the binaural synthesizer and the crosstalk cancellation are affected by the location of the listener's head. The modification to the binaural synthesizer is rather trivial: as the listener moves his head, the synthesis HRTFs must be adjusted so that the rendered scene remains fixed to an external frame of reference, otherwise the rendered scene will move with the listener's head. The head model within the crosstalk canceller must also be updated so that the crosstalk canceller is inverting the current transmission path from the speakers to the ears.

Implementing the head tracking and adaptation has two benefits. First, the 3-D effect will function over a large listening area because the "sweet spot" is steered to the location of the listener's head. Secondly, if the tracking is fast enough, the listener will have the additional benefit of dynamic localization cues. These are very powerful cues, particularly for the resolution of front-back confusions.

3.1.3 Hybrid approach

We will use a hybrid approach to the transmission path inversion problem, using different strategies for low frequencies and high frequencies. This is justified for several reasons:

- The large intersubject variation in high-frequency head response means that it is impossible to invert the head response using a non-individualized head model.

3-D Audio Using Loudspeakers

- Even when using an individualized head model, the high-frequency inversion becomes critically sensitive to positional errors, because the size of the equalization zone is proportional to the wavelength.

- When the cancellation fails due to either a mismatch in head response or positional error, interaural cues may be seriously degraded, and this situation should be avoided.

Therefore, an exact transmission path inversion is only attempted at low frequencies, where intersubject variation in head response is small (Cooper and Bauck, 1990). At high frequencies, a power transfer model is inverted in an attempt to deliver the proper high-frequency powers to each ear. When an exact solution to this model is not possible, the proper total high-frequency power is delivered without regard to interaural cues.

3.1.4 Multichannel implementations

Previous work in multichannel binaural audio using loudspeakers (Yanagida et al., 1983; Abe et al., 1990; Bauck and Cooper, 1992, 1993, 1996) has considered the psuedo-inverse solution to the multichannel problem. When the system is underconstrained (more speakers than ears) this solution minimizes total power sent to the loudspeakers. When the system is overconstrained (fewer speakers than ears), the pseudo-inverse solution minimizes total squared error. We will consider a different approach to the multichannel solution. Our approach is based on the desired property that when a sound source is panned to the location of a loudspeaker, that loudspeaker should emit all the power corresponding to the sound source. We call this the *power panning property*. We note that a binaural synthesizer combined with a two-channel full bandwidth crosstalk canceller will naturally have this property, provided that they use the same head model. This follows directly from the properties of matrix inversion. We can create 3-D audio systems with more than two loudspeakers that have the power panning property by superposing a weighted sum of two-channel crosstalk cancellers (Iwahara and Mori, 1978). The weights are determined by the location of the desired sound source.

3.1.5 Organization of chapter

This chapter begins by describing the head-related transfer function (HRTF) measurements conducted for this study of loudspeaker 3-D audio systems. The HRTF data are discussed, focusing on attributes relevant to this study. This leads into a discussion of the interaural transfer function (ITF), which is the basis of many of our crosstalk canceller designs. The theory of crosstalk cancellation is then presented in detail, including a review of past crosstalk canceller implementations. Methods for bandlimiting the crosstalk cancellation are presented, and this leads to both recursive and non-recursive bandlimited implementations of both symmetric and asymmetric crosstalk cancellers. A high-frequency model is described, analogous to crosstalk cancellation in the usual sense, but expressed in terms of power transfer to the ears. The high-frequency power model is combined with the low-frequency crosstalk cancellation to create a hybrid, frequency dependent approach to crosstalk cancellation.

3.2 HEAD-RELATED TRANSFER FUNCTIONS

Our implementation of the binaural synthesizer and crosstalk canceller are based on head-related transfer functions (HRTFs) measured from a Knowles Electronic Mannequin for Acoustic Research (KEMAR). The KEMAR is an anthropomorphous mannequin consisting of a torso and head with features whose dimensions are based on median human measurements (Burkhard and Sachs, 1975). This section describes the HRTF measurement technique and discusses the resulting measurements. We also discuss several related topics of interest, namely, the equalization of HRTF measurements, invertibility of HRTFs, and modeling of the interaural transfer function. These topics are relevant to the discussion of crosstalk cancellation in the following section.

Strictly speaking, an HRTF is a frequency-domain function which has a corresponding time-domain function called a head-related impulse response (HRIR). An HRTF is obviously related to the HRIR via the Fourier transform. We will use the term HRTF to refer generally to either representation, and will use HRIR to refer specifically to the time-domain representation.

3.2.1 Description of KEMAR HRTF measurements

The HRTF measurements of the KEMAR have been described elsewhere (Gardner and Martin, 1994, 1995). Quoting directly from (Gardner and Martin, 1995):

> Measurements were made using an Apple Macintosh computer equipped with a Digidesign Audiomedia II DSP card, which has 16-bit stereo A/D and D/A converters that operate at a 44.1 kHz sampling rate. One of the audio output channels was sent to an amplifier which drove a Realistic Optimus Pro 7 loudspeaker, a small two-way loudspeaker with a 4 inch woofer and 1 inch tweeter. The KEMAR was Knowles Electronics model DB-4004 and was configured with two neck rings and a torso. The left pinna was the "small" model DB-061 and the right was the "large red" model DB-065. The KEMAR was equipped with Etymotic[†] ER-11 microphones, Etymotic ER-11 preamplifiers, and DB-100 occluded ear simulators with DB-050 ear canal extensions. The outputs of the microphone preamplifiers were connected to the stereo inputs of the Audiomedia card.

[†]Etymotic Research, 61 Martin Lane, Elk Grove Village, IL 60007.

From the standpoint of the Audiomedia card, a signal sent to the audio outputs results in a corresponding signal appearing at the audio inputs. Measuring the impulse response of this system yields the impulse response of the combined system consisting of the Audiomedia D/A and A/D converters and anti-alias filters, the amplifier, the speaker, the room in which the measurements are made, and most importantly, the response of the KEMAR with its associated microphones and preamps. Interference due to room reflections can be avoided by ensuring that any reflections occur well after the head response time, which is several milliseconds.

The measurements were made in MIT's anechoic chamber. The KEMAR was mounted upright on a motorized turntable which could be rotated accurately to any azimuth under computer control. The speaker was mounted on a boom stand which enabled accurate positioning of the speaker to any elevation with respect to the KEMAR. Thus, the measurements were made one elevation at a time, by setting the speaker to the proper elevation and then rotating the KEMAR to each azimuth. With the KEMAR facing forward toward the speaker (0 degrees azimuth), the speaker was positioned such that a normal ray projected from the center of the face of the speaker bisected the interaural axis of the KEMAR at a distance of 1.4 meters. It is believed that the speaker was always within 1.5 cm of the desired position, which corresponds to an angular error of ± 0.5 degrees.

The impulse responses were obtained using a maximum length (ML) sequence measurement technique (Rife and Vanderkooy, 1989; Vanderkooy, 1994). The sequence length was 16383 samples, corresponding to a 14-bit generating register. This sequence length was chosen to yield a good signal to noise ratio (SNR) without excessive storage requirements or computation time. Because the measurements were performed in an anechoic chamber and the ML sequence was sufficiently long, time aliasing in the impulse responses was not significant. The measured SNR for frontal incidence was 65 dB.

The spherical space around the KEMAR was sampled at elevations from -40 degrees (40 degrees below the horizontal plane) to +90 degrees (directly overhead) in 10 degree increments. At each elevation, a full 360 degrees of azimuth was sampled in equal sized increments. The azimuth increment sizes were chosen to maintain approximately 5 degree great-circle increments. Table 3.1 on page 26 shows the number of samples and azimuth increment at each elevation (all angles in degrees). In total, 710 locations were sampled.

It was desired to obtain HRTFs for both the "small" and "large red" pinna styles. If the KEMAR had perfect medial symmetry, including the pinnae, then the resulting set of HRTF measurements would be symmetric within the limits of measurement accuracy. In other words, the left ear response [for a source] at azimuth θ would be equal to the right ear response [for a source] at

Table 3.1 Number of measurements and azimuth increment at each elevation. All angles are in degrees.

Elevation	Number of measurements	Azimuth increment
-40	56	6.43
-30	60	6.00
-20	72	5.00
-10	72	5.00
0	72	5.00
10	72	5.00
20	72	5.00
30	60	6.00
40	56	6.43
50	45	8.00
60	36	10.00
70	24	15.00
80	12	30.00
90	1	n.a.

azimuth 360 - θ. It was decided that an efficient way to obtain symmetrical HRTF measurements for both the "small" and "large red" pinnae was to install both pinnae on the KEMAR simultaneously, and measure the entire 360 degree azimuth circle. This yields a complete set of symmetrical responses for each of the two pinna, by associating each measurement at azimuth θ with the corresponding measurement at azimuth 360 - θ. For example, to form the symmetrical response pair for the "small" pinna (which was mounted on the left ear), given a source location at 45 degrees right azimuth, the left ear response at 45 degrees (contralateral response) would be paired with the left ear response at 315 degrees azimuth (simulated ipsilateral response). Such a symmetrical set will not exhibit interaural differences for sources in the median plane, which has been shown to be a localization cue (Searle et al., 1975). Assuming an HRTF is negligibly affected by the shape of the opposite pinna, these symmetrical sets should be the same as sets obtained using matched pinnae.

3.2.2 Processing

Although a complete set of measurements for both two pinna models were obtained, it was arbitararily decided to use the data from the "small" model DB-061 pinna exclusively for our study of 3-D audio systems. Unless otherwise noted, all KEMAR HRTF data shown in this document are based on a symmetrical set of HRTF data obtained using this pinna model.

3-D Audio Using Loudspeakers

Each ear measurement yielded a 16383 point impulse response at a 44.1 kHz sampling rate. Most of these data are irrelevant. The 1.4 meter air travel corresponds to approximately 180 samples, and there is an additional delay of 50 samples inherent in the playback/recording system. Consequently, in each impulse response, there is a delay of approximately 230 samples followed by the head response, which persists for several hundred samples and is in turn followed by reflections off objects in the anechoic chamber (such as the KEMAR turntable). In order to reduce the size of the data set without eliminating anything of potential interest, the first 200 samples of each impulse response were discarded and the next 512 samples were saved. Each impulse response is thus 512 samples (11.6 msec) long.

Using a frequency-independent window length is problematic. Although the windowing successfully crops trailing echoes, it can also result in the inadvertent loss of low frequencies. The electro-acoustic system being measured is AC-coupled, and its frequency response has a sharp rolloff towards zero at very low frequencies. Windowing the time response of this system is equivalent to computing a circular convolution of the frequency response with the Fourier transform of the window function (Oppenheim and Schafer, 1989). Consequently, applying a short time window has the effect of smoothing the AC-coupling rolloff, and the rolloff towards zero starts at a higher frequency. Using a longer window length reduces the low-frequency smoothing, but also allows additional echoes in the time response, which introduces erroneous details in the frequency response. We note that an approach using wavelet decomposition could allow the time windowing to be longer at low frequencies than at high frequencies, although we have not taken this approach.

The unprocessed measurements contain the frequency response of the measurement system itself, including electronics, Etymotic ER-11 microphone, and most importantly, the Optimus Pro 7 loudspeaker. It is necessary to equalize the data to compensate for the reponse of the measurement system. There are several ways to perform this equalization (Blauert,1983; Moller, 1992; Jot et al., 1995):

1. Equalize the data set with respect to a reference measurement obtained using one of the ear microphones positioned at the center of the head with no head present. This style of equalization gives a clear physical meaning to the equalized measurements: the HRTFs specify the pressure gain at the eardrum relative to the free-field pressure at the center of the head with no head present. We will call this procedure *measurement equalization*.

2. Equalize the data set with respect to an HRTF measurement at a particular direction of incidence. This is called *free-field equalization*[†]. Usually the equalization is done with respect to frontal incidence.

[†]There may be some confusion whether the term "free-field equalized HRTF" should refer to an HRTF that is equalized with respect to free-field incidence at the center of the head with no head present, or to an HRTF equalized with respect to another HRTF at some reference direction. We prefer the latter interpretation, and will use it in this book.

3. Equalize the data set with respect to the diffuse-field average of HRTFs across all incident directions. This is called *diffuse-field equalization*. The diffuse-field average is typically calculated as a power average across a uniformly distributed set of incident locations. Because the power average contains no phase information, the phase response of the equalization filter is arbitrary, and is usually chosen to be minimum phase. According to Blauert (1983), diffuse-field equalization was proposed by Theile in 1981 as an equalization method for dummy head recordings. Theile (1986) also proposed diffuse-field equalization as a standard for headphones. A variant of diffuse-field equalization is to equalize with respect to an average across horizontal locations only, called *horizontal diffuse-field equalization* (Gardner and Martin, 1995).

Each of the equalization methods computes a reference response which is then inverted and used to filter the data set. This operation is often performed in the frequency domain by dividing the complex spectrum of each HRTF with the complex spectrum of the reference response. Issues regarding filter inversion are discussed in Appendix A. Because the equalization response is common to all HRTFs in the data set, ratios of HRTFs are insensitive to the equalization method. Consequently, the free-field and diffuse-field equalizations may be performed on any HRTF data set, regardless of whether the set has been properly measurement equalized. Diffuse-field equalization also requires a uniform spherical distribution of incident directions; horizontal diffuse-field equalization requires a uniform circular distribution of horizontal directions.

In order to perform the measurement equalization for the KEMAR data, the impulse response of the Optimus Pro 7 speaker was measured in free-field conditions in the anechoic chamber at a distance of 1.4 meters. The measurement technique was exactly the same as the HRTF measurements, except that a Neumann[†] KM 84i cardiod microphone was used. This was an oversight that has several ramifications. Equalizing the HRTFs with respect to the Neumann reference measurement does not compensate for any response differences between the Neumann and Etymotic microphones and associated electronics. The KEMAR HRTFs equalized with respect to the Neumann reference have low-frequency phase responses that differ from other published measurements of HRTFs, such as described by Mehrgardt and Mellert (1977) or Moller et al. (1995c). Consequently, we will not use the phase responses of the measurement equalized KEMAR HRTFs.

Almost all of the applications of HRTFs to our study involves forming ratios of HRTFs, either via free-field or diffuse-field equalization, or via the calculation of interaural transfer functions, which are ratios of the two ear responses for a given source direction. Ratios of HRTFs are not affected by the measurement equalization procedure. Discussions of free-field and diffuse-field equalization and interaural

---

[†]Neumann/USA, P.O. Box 987, Old Lyme, CT 06371.

3-D Audio Using Loudspeakers

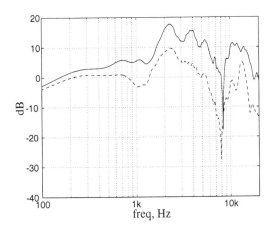

Figure 3.1 KEMAR HRTF magnitude at 30 degrees azimuth on horizontal plane: ipsilateral ear response (solid line) and contralateral ear response (dashed line). The HRTFs are equalized with respect to free-field incidence at the center of the head with no head present. The low-frequency rolloff below 200 Hz is attributed to the time windowing of the HRIRs.

transfer functions are presented in following sections. In this book, an HRTF will refer to a measurement equalized HRTF, i.e., equalized with respect to free-field incidence at the center of the head with no head present, unless it is explicitly stated to be a free-field or diffuse-field equalized HRTF.

3.2.3 Discussion of measurements

Many studies have been made of HRTF measurements of humans and dummy head microphones, for instance, see (Shaw, 1966, 1974; Shaw and Teranishi, 1968; Mehrgardt and Mellert, 1977; Blauert, 1983; Shaw and Vaillancourt, 1985; Wightman and Kistler, 1989a; Moeller et al., 1995c; Larcher and Jot, 1997b). The purpose of discussing our KEMAR measurements in this section is not to provide a complete analysis of the HRTF data, but rather to point out particular details that are relevant to subsequent discussion in this book.

An example HRTF pair is plotted in figure 3.1, which shows the frequency response magnitude of the ipsilateral and contralateral HRTFs for a source at 30 degrees azimuth on the horizontal plane (0 degrees elevation). The responses have many features typical of HRTFs. At low frequencies, the responses are similar, and at higher frequencies the difference in the responses increases, which is attributed to the frequency dependence of head shadowing. The high-frequency responses contain sharp features attributed to interactions of the incident sound with the external ear, for instance, the distinctive notches at 8-9 kHz that are caused by a concha reflection (Lopez-Poveda and Meddis, 1996). The broad peak at 2-3 kHz is caused by the ear canal resonance

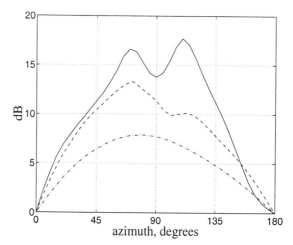

Figure 3.2 Broadband ILDs of KEMAR HRTFs as a function of azimuth angle, at elevation angles of 0 degrees (solid line), 30 degrees (dashed line), and 60 degrees (dash-dot line).

(Blauert, 1983). The shape of these responses are in general agreement with human HRTFs measured by Moller et al. (1995c).

Figure 3.2 shows interaural level differences (ILDs) for the KEMAR HRTF data as a function of azimuth angle, for elevations of 0, 30, and 60 degrees. ILDs were calculated by computing the ratio of the energies of the two HRIRs and converting to dBs; therefore, the ILDs are calculated over the entire frequency range. It should be stressed that the ILD data depends on the equalization of the HRTFs because ILDs are very frequency dependent. The data in figure 3.2 is based on measurement equalized HRTFs, i.e., equalized with respect to free-field incidence at the center of the head with no head present, such as shown in figure 3.1. Because most of the energy in the HRTFs is in the 2-3 kHz band, this band should dominate the ILD data.

At 0 degrees elevation (solid line), the ILD has a characteristic dip at 90 degrees azimuth that is seen in both human HRTF data and spherical head models (Blauert, 1983). Considering a spherical head model, a planar sound wave incident from 90 degrees will diffract around the sphere and will add in phase on the exact opposite side. Consequently, the ILD is less at 90 degrees incidence than it is at 70 degrees incidence, where the contralateral response is a superposition of diffracted waves that have traveled different path lengths and hence are out of phase. The maximum ILD is 17 dB for a horizontal source at 110 degrees azimuth. At 30 degrees elevation, the 90 degree dip is less pronounced, and there is a noticable asymmetry in ILDs for front and rear sources. At 60 degrees elevation, the maximum ILD is about 8 dB. The ILD data characterizes in the most general sense the natural head shadowing that occurs for various source locations. We can use the data to choose loudspeaker placements that maximize natural head shadowing in a loudspeaker 3-D audio system.

Figure 3.3 shows broadband interaural time delays (ITDs) for the KEMAR HRTF data as a function of azimuth angle, for elevations of 0, 30, and 60 degrees. At each elevation, three methods of computing the ITD are compared. The dashed line is the ITD obtained from the KEMAR HRTF data by calculating a linear regression on the excess phase difference of the two ear frequency responses (Jot et al., 1995). The linear regression is calculated over the band 500 Hz - 2 kHz. This procedure is described in more detail in the following section on interaural transfer functions. The dotted line is the ITD obtained from the KEMAR HRTF data by calculating the cross-correlation function of the two ear responses, and equating the ITD to the lag that maximizes the cross-correlation function within ±1 msec. Prior to computing the cross-correlation, the two HRIRs are lowpass filtered with a cutoff frequency of 2 kHz. The solid line shows the ITD calculated for a spherical head model according to (Larcher and Jot, 1997a):

$$ITD = \frac{D}{2c}(\operatorname{asin}(\cos\varphi\sin\theta) + \cos\varphi\sin\theta) \qquad (3.1)$$

where θ is the azimuth angle, φ is the elevation angle, $D = 17.5$ cm is the diameter of the spherical head, and $c = 344$ m/sec is the speed of sound. The data in figure 3.3 shows that the cross-correlation method gives larger ITD estimates than the other methods for near-horizontal sources at extreme lateral azimuths. This agrees with similar results by Larcher[†]. The extent to which these methods are in error depends on how we define the ITD. In the next section we will describe Jot's model for interaural transfer functions which defines the ITD based on the interaural excess phase (Jot et al., 1995).

Figure 3.4a shows frequency response magnitudes of the KEMAR HRTFs as a function of elevation, for medial sources at 0 degrees azimuth. Note that the magnitude spectra are shown with a linear frequency axis. Lightly shaded features in the figure represent spectral peaks. For instance, at 2-3 kHz, the ear canal resonance is visible; the resonance does not change as a function of elevation. Spectral features at frequencies above 5-6 kHz are caused by the filtering effects of the external ear. Spectral peaks in particular are well known to be localization cues (Blauert, 1969/70; Hebrank and Wright, 1974b; Butler and Belendiuk, 1977; Middlebrooks, 1992; Butler, 1997). The spectral peaks do change as a function of elevation. The dependence on elevation is more clearly exhibited by the dark shaded features, which are spectral notches. Spectral notches are caused by multipath reflections off the external ear and have been shown to be elevation cues (Hebrank and Wright, 1974b; Bloom, 1977). Regardless of the relative salience of the notches and peaks as localization cues, the notches are compelling because they reveal the underlying physics of the external ear. In Lopez-Poveda's study of the physics of the external ear (Lopez-Poveda, 1996; Lopez-Poveda and Meddis, 1996), the three notches in figure 3.4a were described as

---

[†]Véronique Larcher, personal communication, 1997.

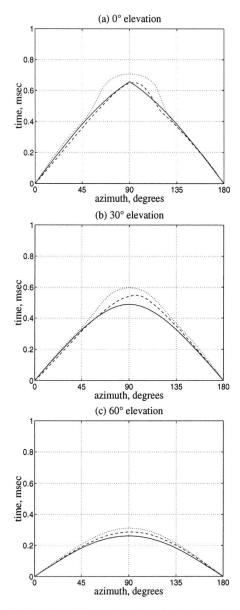

Figure 3.3 ITDs of KEMAR HRTFs as a function of azimuth angle, at elevation angles of 0 degrees (a), 30 degrees (b), and 60 degrees (c). Dashed line shows KEMAR ITD calculated by linear regression on the interaural excess phase, dotted line shows KEMAR ITD calculated by cross-correlation, and solid line shows ITD calculated from a spherical head model.

3-D Audio Using Loudspeakers

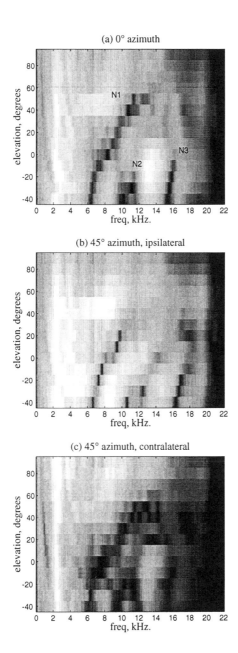

Figure 3.4 Magnitude spectra of KEMAR HRTFs as a function of elevation: source at 0 degrees azimuth (a); source at 45 degrees azimuth, ipsilateral ear (b); source at 45 degrees azimuth, contralateral ear (c). White indicates +10 dB, black indicates -30 dB. Notch features are labeled in (a) according to Lopez-Poveda and Meddis (1996).

N1, N2, and N3, and are so labeled in the figure. The notch N1, attributed to a relection off the posterior wall of the concha (Hebrank and Wright, 1974b), ranges from 6–12 kHz for source elevations of -40 to 60 degrees. Lopez-Poveda and Meddis (1996) also attributed the N3 notch to a concha interaction, but were not able to describe N2 in terms of concha physics. Lopez-Poveda (1996) found similar spectral features in the frequency responses of both KEMAR and human HRTFs.

Figures 3.4b and 3.4c show the KEMAR HRTF spectra as a function of elevation for a source at 45 degrees: the ipsilateral ear spectra are shown in figure 3.4b, and the contralateral ear spectra are shown in figure 3.4c. Again the same notches are seen, with the same dependency on elevation, although the notches are less distinct in the contralateral ear response. We note that in all three figures there is relatively little spectral variation at elevations greater than 60 degrees, except for the ubiquitous ear canal resonance at 2-3 kHz.

Figure 3.5 shows the KEMAR HRTF magnitude spectra as a function of source azimuth for horizontal sources: figure 3.5a shows the ipsilateral ear response, and figure 3.5b shows the contralateral ear response. Together, the two figures show the responses for a full 360 degrees of source azimuths. In figure 3.5a, the distinctive notch N1 is quite apparent. It ranges from 8-10 kHz for ipsilateral source locations. It is remarkable that the N1 feature has a dependence on elevation that is largely independent of azimuth, a phenomenon discussed in detail by Lopez-Poveda (1996). The labeling of the N2 feature is speculative on our part; at 45 degrees azimuth the notch is at 14 kHz, and this matches the N2 notch in figure 3.4b at 0 degrees elevation. In the contralateral response (figure 3.5b) at frequencies below 6 kHz. there is a regular pattern of peaks and dips that is both frequency and azimuth dependent. We attribute this to the reinforcement and cancellation of waves travelling different paths around the head.

3.2.4 Equalization of HRTFs

Our implementations of binaural synthesizers discussed in this chapter and following chapters exclusively use either diffuse-field or free-field equalized HRTFs, depending upon the context. The diffuse-field average is formed as the power average across all locations, according to:

$$\left|H_{\text{DF}}(e^{j\omega})\right| = \sqrt{\frac{1}{N}\sum_{i=1}^{N}\left|H_i(e^{j\omega})\right|^2} \tag{3.2}$$

The power averaging establishes the magnitude of the diffuse-field response; the phase is left unspecified. The diffuse-field average of the unprocessed KEMAR HRTF measurements is shown in figure 3.6. It was obtained by evaluating equation 3.2, and then smoothing the result by applying a 1/3-octave constant-Q

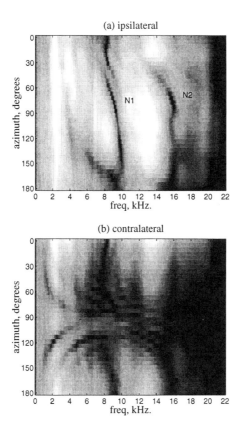

Figure 3.5 Magnitude spectra of KEMAR HRTFs as a function of azimuth for a horizontal source: ipsilateral ear (a), contralateral ear (b). White indicates +10 dB, black indicates -30 dB. Notch features are labeled in (a) according to Lopez-Poveda and Meddis (1996)

smoothing filter. We have adopted the practice of smoothing filter responses used for HRTF equalization in order to simplify the resulting filter response; in this particular case the smoothing had little effect because the diffuse-field average response was already smooth. The diffuse-field response contains all non-directional components of the measurements, including the measuring system and ear canal resonance. The diffuse-field response was inverted to create a diffuse-field equalization filter that was applied to all the HRTFs in the set. The phase response of the inverse filter was determined using the Hilbert transform, such that the inverse filter was a minimum-phase function (Oppenheim and Schafer, 1989).

Free-field equalized HRTFs are created by equalizing the set of HRTFs with respect to an HRTF measured at one ear for a particular direction of sound incidence. For reasons that are explained in following sections, we will often use HRTFs that are equalized with respect to the ipsilateral response at 30 degrees horizontal incidence, a

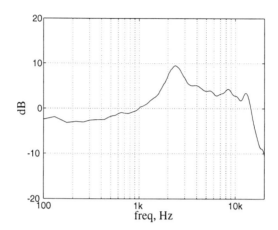

Figure 3.6 Diffuse-field average of KEMAR HRTFs, after smoothing with a 1/3-octave constant-Q smoothing filter.

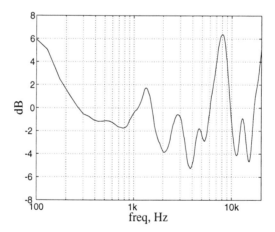

Figure 3.7 30 degree free-field equalization response, derived from diffuse-field equalized HRTFs. The equalization response has been smoothed with a 1/3-octave constant-Q smoothing filter.

direction that corresponds to the speaker location in a typical listening arrangement. Figure 3.7 shows a 30 degree free-field equalization response obtained from the diffuse-field equalized HRTF at 30 degrees incidence. The equalization filter was calculated by smoothing the HRTF magnitude response using a 1/3-octave constant-Q smoothing filter and then inverting the smoothed reponse. Had the smoothing not been done, the inverse filter would have sharp high-frequency peaks corresponding to the notches shown in figure 3.1 on page 29. The equalization filter in figure 3.7 was used to create a set of 30 degree free-field equalized HRTFs by filtering each diffuse-field equalized HRTF with the free-field equalization filter.

3-D Audio Using Loudspeakers

The low-frequency boost is an abberation caused by the windowing phenomenon discussed earlier. Each diffuse-field equalized HRTF should be reasonably flat at low frequencies, but because they are AC-coupled and windowed to 128 points (at 44.1 kHz sampling rate), they suffer from poor low-frequency response. Consequently, the free-field equalization filter has a boost to compensate. We have recently begun artificially restoring the DC component of the unprocessed HRTFs prior to further processing. This technique greatly reduces low-frequency loss due to windowing.

The free-field equalization filter has a number of peaks in its response. Of particular note are the peaks at 1.5 kHz and 8 kHz. These correspond to the "boosted bands" noted by Blauert (1969/70) for which sound pressure at the ear in a majority of human subjects was greater for rear sources than for frontal sources. Thus, KEMAR exhibits the same front-back spectral differences as seen in a majority of human subjects. Blauert (1969/70) showed that these rear boosted bands correspond to "directional bands" for which the majority of subjects chose the rear rather than front direction when presented with 1/3-octave noise stimuli at these frequencies. The KEMAR diffuse-field equalized response at 30 degrees (frontal) incidence has a deficiency of energy in these bands. Equalizing with respect to 30 degrees incidence therefore has the effect of applying spectral cues that should favor rear localization. It should be cautioned, however, that these types of spectral cues are seen only for majorities of subjects; we expect significant individual variation in high-frequency spectral cues. Furthermore, in Blauert's study, the band at 8 kHz was principally associated with the overhead direction.

3.2.5 The interaural transfer function

The diffraction of sound by the head, or head shadowing, can be described for a given source location by computing the ratio of the frequency responses at the two ears. This ratio is called the interaural transfer function (ITF). ITFs are important to study because they describe crosstalk. The usual definition of the ITF (e.g., Blauert, 1983) is:

$$ITF = \frac{H_c}{H_i} \qquad (3.3)$$

where H_c is the contralateral response and H_i is the ipsilateral response, expressed in the frequency domain. In this section, we will use this convention of referencing the ITF to the same side response so that the associated ITD is positive. It should be clear that either ratio of the two ear responses is a valid ITF. Figure 3.8 shows the magnitude response of the KEMAR ITF at 30 degrees horizontal incidence. This is calculated by dividing the contralateral magnitude response with the ipsilateral magnitude response (see figure 3.1). At frequencies below 6 kHz, the ITF behaves like a lowpass filter with a gentle rolloff, but at higher frequencies the ITF magnitude has large peaks corresponding to notches in the ipsilateral response. Figure 3.9 shows the interaural

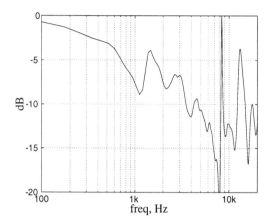

Figure 3.8 Interaural transfer function (ITF) magnitude at 30 degrees horizontal incidence, derived from KEMAR HRTFs. This function describes head shadowing for a 30 degree horizontal source. The function was smoothed with a 1/24th-octave constant-Q smoothing filter.

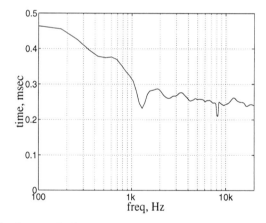

Figure 3.9 Interaural phase delay of KEMAR HRTFs at 30 degree horizontal incidence. The ITD, calculated by cross-correlation, is 0.25 msec. The increase in phase delay at frequencies below 1.5 kHz is predicted by the theory of sound diffraction by a sphere (Rayleigh, 1945; Kuhn, 1977, 1987).

phase delay at 30 degrees incidence. The interaural phase delay is the difference of the unwrapped phases of the two ear responses, divided by the angular frequency. For comparison, the ITD, determined by cross-correlation, is 0.25 msec. The increase in interaural phase delay below 1.5 kHz seen in figure 3.9 is predicted by the theory of sound diffraction by a sphere (Rayleigh, 1945; Kuhn, 1977, 1987).

Calculation of the ITF is accomplished by a convolution of the contralateral response with the inverse ipsilateral response. The causality and stability of the interaural transfer function depends on the invertibility of the ipsilateral HRIR. We now consider in general the invertibility of HRIRs. As discussed in Appendix A, a finite length impulse response can be inverted using a causal and stable filter if and only if the original impulse response is a minimum-phase function. Thus, calculation of equation 3.3 results in a causal and stable ITF if and only if the ipsilateral HRIR is a minimum-phase function.

Mehrgardt and Mellert (1977) suggested that HRIRs are minimum-phase functions. Subsequent research has shown that HRIRs often contain non-minimum-phase zeros at high frequencies, typically above 10 kHz (Moller et al., 1995c; Jot et al., 1995). Our own measurements of the KEMAR HRTFs also indicate that non-minimum phase zeroes occur at high frequencies. A non-minimum phase acoustic impulse response can result from delayed reflections that are more energetic than the direct response. It is easy to imagine that at high frequencies the external ear creates focused reflections that are more intense than the first wavefront, resulting in non-minimum-phase zeros at high frequencies. Because HRIRs are in general non-minimum-phase functions, an inverse HRIR filter that is stable must also be anticausal (see Appendix A). An inverse filter for an HRIR is in general anticausal and has infinite time support.

Because the sound wavefront reaches the ipsilateral ear first, it is tempting to think that the ITF has a causal time representation. However, the true inverse ipsilateral response will be infinite and two-sided because of non-minimum-phase zeros in the ipsilateral response. Therefore the ITF will also have infinite and two-sided time support. Nevertheless, it is possible to accurately approximate the ITF at low frequencies using causal (and stable) filters. Causal implementations of ITFs are required to implement realtime filters that can model head shadowing.

Any rational system function can be decomposed into a minimum-phase system cascaded with an allpass-phase system (Oppenheim and Schafer, 1989), which we notate as:

$$H(z) = \text{minp}(H(z))\text{allp}(H(z)) \qquad (3.4)$$

The ITF can then be expressed as the ratio of the minimum phase systems cascaded with an allpass system whose phase response is the difference of the excess (allpass) phases at the two ears:

$$ITF(e^{j\omega}) = \frac{\text{minp}(H_c(e^{j\omega}))}{\text{minp}(H_i(e^{j\omega}))} e^{j(\angle\text{allp}(H_c(e^{j\omega})) - \angle\text{allp}(H_i(e^{j\omega})))} \qquad (3.5)$$

Jot has shown that for all incidence angles, the excess phase difference in equation 3.5 is approximately linear with frequency at low frequencies. Therefore the ITF can be modeled as a frequency-independent delay cascaded with the minimum phase part of the true ITF (Jot et al., 1995):

$$ITF(e^{j\omega}) \cong \frac{\mathrm{minp}(H_c(e^{j\omega}))}{\mathrm{minp}(H_i(e^{j\omega}))} e^{-j\omega ITD/T} \tag{3.6}$$

where *ITD* is the frequency-independent interaural time delay, and T is the sampling period. This ITF model is stable and causal, which is of paramount importance for realtime implementation. The ITD is optimally calculated by linear regression on the interaural excess phase term given in equation 3.5 (Jot et al., 1995). The regression may be calculated over a low-frequency band, such as 500 Hz - 2 kHz. It is important to note the relationship between the *excess* phase difference, which leads to a frequency-independent ITD, and the *total* phase difference, which leads to an interaural phase delay that increases at frequencies below 1.5 kHz, as shown in figure 3.9.

We have determined that the model in equation 3.6 is accurate at low frequencies for any ratio of two HRTFs, and not just interaural transfer functions. This follows directly from the fact that non-minimum-phase zeros in HRTFs only occur at high frequencies. The excess phase part of an HRTF therefore consists of a linear phase term and one or more second-order allpass sections at high frequencies. Consequently, the excess phase difference between any two HRTFs must be nearly linear at low frequencies, where the excess phase responses are negligibly affected by the high-frequency allpass resonances.

The crosstalk cancellers we discuss in later sections require the implementation of lowpass filtered ITFs. Because the ITFs are used to model acoustic crosstalk for cancellation, accurate phase response is critical. This suggests that we filter the ITF with a zero-phase lowpass filter. However, this conflicts with the goal of a causal ITF. The solution is to steal m samples of modeling delay from the ITD in order to design a lowpass filter that is approximately (or exactly) linear phase with a phase delay of m samples. After lowpass filtering the ITF, we can extract a causal filter that models head shadowing at low frequencies. We generalize the approximation to the lowpass filtered ITF as follows:

$$H_{LPF}(e^{j\omega})ITF(e^{j\omega}) \cong L(e^{j\omega})e^{-j\omega(ITD/T - m)}$$

$$\text{s.t.}$$
$$l[n] = 0, \quad n < 0 \tag{3.7}$$
$$\angle H_{LPF}(e^{j\omega}) \cong -m\omega$$

where $l[n]$ is a causal filter with frequency response $L(e^{j\omega})$ that approximates the lowpass filtered ITF after delaying by $ITD/T - m$ samples, and m is the modeling delay of $H_{LPF}(e^{j\omega})$ taken from the ITD. The closest approximation is obtained when all the available ITD is used for modeling delay. However, we may want a parameterized implementation which cascades a filter $L(z)$ with a variable delay to simulate an azimuth dependent ITF. In this case the range of the simulated azimuths is increased if we minimize m.

There are two approaches to obtaining the filter $L(z)$. The first is to start with Jot's ITF model of equation 3.6, which entails 1) separating the HRTFs into minimum-phase and excess phase parts, 2) estimating the ITD by linear regression on the interaural excess phase, 3) computing the minimum phase ITF, and 4) delaying this by the estimated ITD. Figure 3.10a shows the result of this procedure, for an ITF at 30 degree horizontal incidence. The filter $L(z)$ can then be obtained by lowpass filtering and extracting $l[n]$ from the time response starting at sample index $floor(ITD/T - m)$.

Our approach, first presented by Gardner (Casey et al., 1995)[†], has been to compute the ITF by convolving the contralateral response with an inverse ipsilateral response computed using the DFT procedure described in Appendix A. This yields a two-sided time response whose anticausal portion contains high-frequency ringing attributed to the non-minimum phase zeros in the ipsilateral response. Figure 3.10b shows the result of this procedure for an ITF at 30 degree horizontal incidence. After lowpass filtering, the anticausal part of the response will be greatly attenuated. As with the previous procedure, the filter $L(z)$ is obtained by lowpass filtering, rectangular windowing and time shifting.

Figure 3.10c shows the results of both procedures after lowpass filtering with a zero-phase FIR filter with a 6 kHz cutoff. The time responses are very similar. Note that both responses contain ringing of the lowpass filter at negative time because we have not yet windowed the responses to extract causal filters. We can evaluate an ITF model by computing the following error as a function of frequency:

$$\varepsilon = 20\log_{10}\left(\frac{|ITF_{model} - ITF_{ideal}|}{|ITF_{ideal}|}\right) \tag{3.8}$$

This is the error between the model and ideal ITFs, normalized by the magnitude of the ideal ITF. Figure 3.10d shows this error for each of the two ITF calculation methods, computed for 256-pt FIR filters extracted from the responses in figure 3.10c

---

[†]The method of computing lowpass filtered ITFs was presented at the conference, but not included in the proceedings.

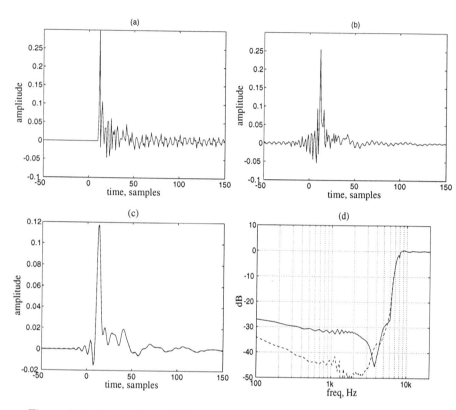

Figure 3.10 Comparison of ITF calculation methods for KEMAR HRTFs at 30 degrees horizontal incidence. (a) ITF time response calculated using minimum-phase model, 44.1 kHz sampling rate. (b) ITF time response calculated using anticausal ipsilateral inverse filter. (c) The two responses overlayed after lowpass filtering at 6 kHz cutoff: solid line is minimum-phase method, dashed line is anticausal method. The two responses are nearly identical. By windowing either response to extract positive time samples, a causal, FIR filter that approximates the lowpass filtered ITF can be obtained. (d) Error comparison of responses in (c) after extracting 256-pt FIR filters starting at sample index 0: solid line is minimum-phase method, dashed line is anticausal method.

starting at sample index 0. The method inspired by Jot's model has slightly more error because of the errors in modeling the excess phase difference as a constant delay, and estimating and synthesizing this delay. Otherwise the two methods are extremely similar.

Implementing the lowpass filtered ITFs using recursive IIR filters will enable significant computation savings. A complete discussion of techniques used to design IIR filters is beyond the scope of this document, but a few comments are in order. An important point is that most of the optimal filter design techniques work by minimizing the error in the domain of frequency response magnitude. These techniques are

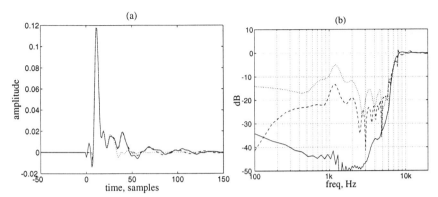

Figure 3.11 Comparison of FIR and IIR implementations of lowpass filtered ITFs, for 30 degree horizontal incidence: time responses at 44.1 kHz (a); error as a function of frequency (b). Solid line is 256-point FIR filter starting at sample index 0, dashed line is 16th-order IIR filter, and dotted line is 8th-order IIR filter. Both IIR responses in (a) start at sample index 7 (the modeling delay m is 4 samples and the ITD is 11 samples).

insensitive to phase and generally result in the design of a minimum-phase filter. An example is the Yule-Walker method (Friedlander and Porat, 1984). These techniques can be used to design an IIR approximation to the minimum-phase part of the ITF. We can then cascade this minimum-phase filter with a delay and a lowpass filter to obtain the lowpass ITF model of equation 3.6. A problem with this method is that the filter design procedure will allocate most of the filter poles and zeros to approximating high-frequency features that are subsequently filtered by the lowpass. This can be alleviated somewhat by using a warped filter design method, e.g. the Bark bilinear transform (Smith and Abel, 1995), or by weighting the error criteria so that only the low-frequency portion of the response is approximated.

Our approach has been to use Prony's method (Weiss and McDonough, 1963; Burrus and Parks, 1970, 1987) to approximate the filter impulse response $l[n]$. Prony's method minimizes the squared error in the time domain, and is therefore sensitive to phase. Also, Prony's method naturally allocates filter poles to approximating high energy features. Because $l[n]$ is lowpass filtered, the time response is very simple and can be accurately modeled with low-order IIR filters. Figure 3.11a compares the time responses of lowpass ITFs modeled with FIR and IIR filters. The solid line is the causal FIR reponse obtained using the anticausal inverse method, i.e. the same as the solid line in figure 3.10c, but after windowing positive time samples. The dashed and dash-dot lines are 16th and 8th-order IIR filter responses, respectively, designed using Prony's method. For these IIR filters, the modeling delay m is 4 samples, and the responses are delayed by 7 samples to obtain the correct ITD of 11 samples, according to equation 3.7. Figure 3.11b shows the errors for each of the three filters. The IIR filters have significantly more error than the 256-point FIR filter, and as expected the 16th-order filter is a better approximation than the 8th-order filter. Nev-

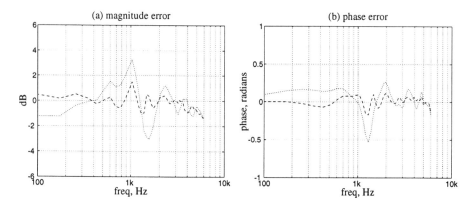

Figure 3.12 Magnitude (a) and phase (b) errors between bandlimited IIR models and the KEMAR ITF: dashed line is error for the 16th-order filter, dotted line is the error for the 8th-order filter.

ertheless, the 8th-order IIR filter is quite adequate for implementing crosstalk cancellers.

Figure 3.12 shows the magnitude error (a) and phase error (b) between the IIR models and the KEMAR ITF: the dashed line is the 16th-order filter, and the dotted line is the 8th-order filter. As expected, the 16th-order filter models both the phase delay and the magnitude better than the 8th-order filter. A comparison of figure 3.12 and figure 3.11b reveals that the largest ITF modeling errors are attributable to the phase errors, and not the magnitude errors, in the low order models. Both the 8th and 16th-order filters have the largest phase errors between 1 and 2 kHz where the KEMAR interaural phase delay has a pronounced dip as shown in figure 3.9.

Figure 3.13 plots the lowpass filtered ITF magnitudes for horizontal sources from 5 to 45 degrees azimuth. The ITFs share a number of similar features in the form of local minima and maxima in the magnitude responses. The IIR design method will assign filter poles and zeros to reproduce these features. Because the features change slowly as a function of azimuth, so will the corresponding poles and zeros. Thus it seems reasonable to expect that intermediate ITFs can be approximated well by interpolating the IIR filter coefficients of two adjacent ITFs. Methods for interpolating IIR filter coefficients are beyond the scope of this book.

3.3 THEORY OF CROSSTALK CANCELLATION

This section will present the theory of crosstalk cancellation, including a review of previous work and some new theory and insights original to this work. Many of the equations deal with linear systems, and the transfer functions that relate input and output signals. In order that convolution be expressed as a multiplication, these equations are expressed in the frequency domain, and for simplicity the frequency

3-D Audio Using Loudspeakers

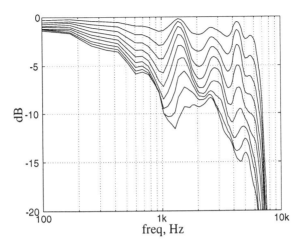

Figure 3.13 KEMAR ITF magnitudes for horizontal source azimuths from 5 to 45 degrees in 5 degree increments (top to bottom plot, respectively), lowpass filtered with a 6 kHz cutoff. The ITFs were smoothed using a 1/3 octave constant-Q smoothing filter.

variables are omitted wherever possible. Unless otherwise stated, all signals are frequency domain representations of their time domain counterparts. Scalar signals are notated in lower case, transfer functions in upper case. Vectors and matrices are both notated using boldface, vectors in lower case, and matrices in upper case.

3.3.1 Binaural synthesis

Binaural synthesis is accomplished by convolving an input signal with a pair of HRTFs:

$$\mathbf{x} = \mathbf{h}x$$

$$\mathbf{x} = \begin{bmatrix} x_L \\ x_R \end{bmatrix}, \mathbf{h} = \begin{bmatrix} H_L \\ H_R \end{bmatrix} \tag{3.9}$$

where x is the input signal, $\mathbf{x}$ is a column vector of binaural signals, and $\mathbf{h}$ is a column vector of synthesis HRTFs. This is a general specification of the binaural synthesis procedure; there are many efficient ways to implement the synthesis filters (Jot et al., 1995). We call the vector $\mathbf{x}$ a *binaural signal* because it would be suitable for headphone listening, perhaps with some additional equalization applied.

The binaural signal may be a sum of multiple input sounds rendered at different locations:

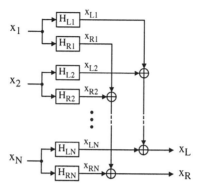

Figure 3.14 Multiple source binaural synthesizer.

$$\mathbf{x} = \sum_{i=1}^{N} \mathbf{h}_i x_i \qquad (3.10)$$

where $\mathbf{h}_i$ is the HRTF vector for source x_i. Figure 3.14 shows the circuit that implements the multiple source binaural synthesizer. For simplicity, in the ensuing discussion the binaural synthesis procedure will be specified for a single source only.

When the binaural signal is being reproduced, rather than synthesized, the individual signals will have been recorded with spatial cues encoded, in which case the synthesis HRTFs have already been applied. Using a prerecorded binaural signal constrains the subsequent processing that can be done because it is not possible to manipulate the individual synthesis HRTFs without first performing a complicated unmixing procedure.

3.3.2 General asymmetric crosstalk cancellation

In order to deliver the binaural signal over loudspeakers, it is necessary to filter it appropriately with a 2x2 matrix $\mathbf{C}$ of transfer functions:

$$\mathbf{y} = \mathbf{C}\mathbf{x}$$

$$\mathbf{y} = \begin{bmatrix} y_L \\ y_R \end{bmatrix}, \mathbf{C} = \begin{bmatrix} C_{11} & C_{12} \\ C_{21} & C_{22} \end{bmatrix} \qquad (3.11)$$

We will call the vector of loudspeaker signals $\mathbf{y}$ a *loudspeaker binaural signal*, and the filter $\mathbf{C}$ the *crosstalk canceller*. Because much of our discussion will concern different implementations of the crosstalk canceller, we have chosen $\mathbf{x}$ and $\mathbf{y}$ to be the input and output variables.

3-D Audio Using Loudspeakers

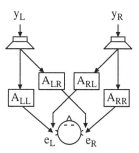

Figure 3.15 Acoustic transfer functions between two loudspeakers and the ears of a listener.

The standard two channel listening situation is depicted in figure 3.15. The ear signals are related to the speaker signals through the equation:

$$\mathbf{e} = \mathbf{A}\mathbf{y}$$

$$\mathbf{e} = \begin{bmatrix} e_L \\ e_R \end{bmatrix}, \mathbf{A} = \begin{bmatrix} A_{LL} & A_{RL} \\ A_{LR} & A_{RR} \end{bmatrix} \quad (3.12)$$

where $\mathbf{e}$ is a column vector of ear signals, $\mathbf{A}$ is the acoustical transfer matrix, and $\mathbf{y}$ is a column vector of speaker signals. The ear signals are considered to be measured by an ideal transducer somewhere in the ear canal such that all direction-dependent features of the head response are captured. The functions A_{XY} give the transfer function from speaker $X \in \{L, R\}$ to ear $Y \in \{L, R\}$ and include the speaker frequency response, air propagation, and head response. $\mathbf{A}$ can be factored as follows:

$$\mathbf{A} = \mathbf{HS}$$

$$\mathbf{H} = \begin{bmatrix} H_{LL} & H_{RL} \\ H_{LR} & H_{RR} \end{bmatrix}, \mathbf{S} = \begin{bmatrix} S_L A_L & 0 \\ 0 & S_R A_R \end{bmatrix} \quad (3.13)$$

$\mathbf{H}$ is the *head transfer matrix* which is a matrix of HRTFs normalized with respect to the free-field response at the center of the head, with no head present. The measurement point of the HRTFs, for example at the entrance of the ear canal, and hence the definition of the ear signals $\mathbf{e}$, is left unspecified to simplify the discussion. $\mathbf{S}$ is the *speaker and air transfer matrix* which is a diagonal matrix that accounts for the frequency response of the speakers and the air propagation to the listener. S_X is the frequency response of speaker X and A_X is the transfer function of the air propagation from speaker X to the center of the head, with no head present. A simplifying assumption is that each speaker response S_X affects the ipsilateral and contralateral ears equally.

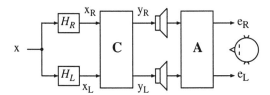

Figure 3.16 Schematic of playback system including binaural synthesizer, crosstalk canceller, and acoustic transfer to the listener.

The playback system is shown in figure 3.16. In order to exactly deliver the binaural signals to the ears, the crosstalk canceller **C** is chosen to be the inverse of the acoustical transfer matrix:

$$\mathbf{C} = \mathbf{A}^{-1} = \mathbf{S}^{-1}\mathbf{H}^{-1} \qquad (3.14)$$

This implements the transmission path inversion. $\mathbf{H}^{-1}$ is the inverse head transfer matrix, later discussed in detail. $\mathbf{S}^{-1}$ associates an inverse filter with each speaker output:

$$\mathbf{S}^{-1} = \begin{bmatrix} 1/(S_L A_L) & 0 \\ 0 & 1/(S_R A_R) \end{bmatrix} \qquad (3.15)$$

The $1/S_X$ terms invert the speaker frequency responses and the $1/A_X$ terms invert the air propagation. In practice, this equalization stage may be omitted if the listener is equidistant from two well-matched, high quality loudspeakers. However, when the listener is off axis, it is necessary to delay and attenuate the closer loudspeaker so that the signals from the two loudspeakers arrive simultaneously at the listener and with equal amplitude. This signal alignment is accomplished by the $1/A_X$ terms above.

In a realtime implementation, it is necessary to cascade the crosstalk canceller with enough modeling delay to create a causal system. Adding a discrete-time modeling delay of m samples to equation 3.14, we obtain:

$$\mathbf{C}(z) = z^{-m}\mathbf{S}^{-1}(z)\mathbf{H}^{-1}(z) \qquad (3.16)$$

The amount of modeling delay needed will depend on the particular implementation. In order to simplify the following discussion, we will omit the modeling delay and the speaker equalization $\mathbf{S}^{-1}$ terms, and consider only the inverse head transfer matrix. Thus, while we recognize that equation 3.14 is the general solution, we will consider crosstalk cancellers of the form:

3-D Audio Using Loudspeakers

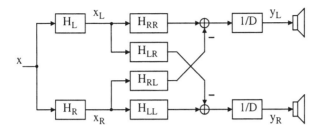

Figure 3.17 Single source binaural synthesizer cascaded with crosstalk cancellation filter. The crosstalk cancellation filter is implemented using four feedforward filters and two inverse determinant filters, where $D = H_{LL}H_{RR} - H_{LR}H_{RL}$. The symmetric form of this circuit was first described by Schroeder and Atal (1963). In their implementation the inverse determinant filter was commuted to the input of the circuit before the binaural synthesis stage.

$$\mathbf{C} = \mathbf{H}^{-1} \tag{3.17}$$

We will use the general form of equation 3.14 whenever the complete playback system is discussed. The inverse head transfer matrix is:

$$\mathbf{H}^{-1} = \begin{bmatrix} H_{RR} & -H_{RL} \\ -H_{LR} & H_{LL} \end{bmatrix} \frac{1}{D} \tag{3.18}$$

$$D = H_{LL}H_{RR} - H_{LR}H_{RL}$$

where D is the determinant of the matrix $\mathbf{H}$. The inverse determinant $1/D$ is common to all terms and determines the stability of the inverse filter. However, because it is a common factor, it only affects the overall equalization and does not affect crosstalk cancellation. When the determinant is 0 at any frequency, the head transfer matrix is singular and the inverse matrix is undefined.

Figure 3.17 shows the schematic of a single source binaural synthesizer and the crosstalk canceller of equation 3.17. This flow diagram was described by Schroeder and Atal (1963). In their implementation, the inverse determinant filter was commuted to the input of the circuit before the binaural synthesis stage.

Dividing numerator and denominator by $H_{LL}H_{RR}$, equation 3.18 can be rewritten as (Moller, 1992):

$$\mathbf{H}^{-1} = \begin{bmatrix} 1/H_{LL} & 0 \\ 0 & 1/H_{RR} \end{bmatrix} \begin{bmatrix} 1 & -ITF_R \\ -ITF_L & 1 \end{bmatrix} \frac{1}{1 - ITF_L ITF_R} \tag{3.19}$$

where

$$ITF_L = \frac{H_{LR}}{H_{LL}}, ITF_R = \frac{H_{RL}}{H_{RR}} \qquad (3.20)$$

are the interaural transfer functions (ITFs). An examination of equation 3.19 reveals much about the crosstalk cancellation process. Crosstalk cancellation is effected by the $-ITF$ terms in the off-diagonal positions of the righthand matrix. These terms predict the crosstalk and send an out-of-phase cancellation signal into the opposite channel. For instance, the right input signal is convolved with ITF_R, which predicts the crosstalk that will reach the left ear, and the result is subtracted from the left output signal. The common term $1/(1 - ITF_L ITF_R)$ compensates for higher-order crosstalks, in other words the fact that each crosstalk cancellation signal itself transits to the opposite ear and must be cancelled. It is a power series in the product of the left and right interaural transfer functions, which explains why both ear signals require the same equalization signal: both ears receive the same high-order crosstalks. Because crosstalk is more significant at low frequencies, this term is essentially a bass boost. The lefthand diagonal matrix, which we call *ipsilateral equalization*, associates the ipsilateral inverse filter $1/H_{LL}$ with the left output and $1/H_{RR}$ with the right output. These are essentially high-frequency spectral equalizers that facilitate the perception of rear sources using frontal loudspeakers (see "Equalization of HRTFs" on page 34). The use of the ITF to predict crosstalk at the contralateral ear requires that each output be equalized with respect to ipsilateral incidence. The ipsilateral equalization filters also compensate for any asymmetries in path lengths from speakers to ears when the head is rotated.

Using equation 3.19, the transfer functions for the circuit of figure 3.17 can be written as (Moller, 1992):

$$\begin{bmatrix} y_L/x \\ y_R/x \end{bmatrix} = \begin{bmatrix} \left(\frac{H_L}{H_{LL}}\right) - \left(\frac{H_R}{H_{LL}}\right) ITF_R \\ \left(\frac{H_R}{H_{RR}}\right) - \left(\frac{H_L}{H_{RR}}\right) ITF_L \end{bmatrix} \frac{1}{1 - ITF_L ITF_R} \qquad (3.21)$$

An examination of equation 3.21 reveals that it is composed entirely of ratios of HRTFs which correspond to either ITFs or free-field equalized HRTFs. This is an important point, because it means that any factor common to the HRTFs will cancel. Thus, the HRTFs can be measured at any location within the ear canal or at the entrance of the blocked ear canal. Similarly, the HRTFs may be free-field or diffuse-field equalized. All of these possibilities yield the same solution. The only constraint is that the HRTFs used for the binaural synthesizer be equalized the same as the HRTFs used for the crosstalk canceller.

3-D Audio Using Loudspeakers

In practice, the listener's HRTFs may not be exactly equal to the head model used by the crosstalk canceller. In this case, the condition for perfect crosstalk cancellation is that the matrix $\mathbf{HM}^{-1}$ be diagonal, where $\mathbf{H}$ is the true head transfer matrix, and $\mathbf{M}$ is an analogous matrix of model head transfer functions. The matrix $\mathbf{HM}^{-1}$ is diagonal when

$$\frac{H_{RL}}{H_{LL}} = \frac{M_{RL}}{M_{LL}}$$
$$\text{and} \qquad\qquad\qquad (3.22)$$
$$\frac{H_{LR}}{H_{RR}} = \frac{M_{LR}}{M_{RR}}$$

These ratios are not ITFs and they don't have an intuitive physical interpretation.

3.3.3 Stability and realizability

The matrix $\mathbf{H}$ is invertible if and only if it is non-singular, i.e. if its determinant $D \neq 0$ (see equation 3.18). Because $\mathbf{H}$ is a function of frequency, it is possible that the inverse matrix $\mathbf{H}^{-1}$ exists only for particular frequency ranges where the matrix $\mathbf{H}$ is non-singular. Similarly, if the matrix $\mathbf{H}$ is poorly conditioned at some frequency, this will lead to a small value of D, and the magnitude of $1/D$ will be very large. In practice, this can be handled by limiting the magnitude of $1/D$, and in these frequency ranges the inverse matrix only approximates the true inverse.

The form of the inverse matrix given in equation 3.19 is obtained by dividing by H_{LL}-H_{RR}. Thus, an additional constraint for the existence of this form is that $H_{LL} \neq 0$ and $H_{RR} \neq 0$. The inverse ipsilateral filters and the interaural transfer functions both depend on this constraint. As before, we may limit the magnitude of $1/H_{LL}$ and $1/H_{RR}$ in order to obtain approximate inverses in frequency ranges where the magnitudes of the ipsilateral transfer functions are small.

Our goal is to implement the crosstalk canceller using realtime digital signal processing methods, and this implies a causal, discrete-time implementation. For the present discussion we will also assume that the crosstalk canceller is a linear, time-invariant (LTI) system[†]. We consider LTI systems whose z-transforms can be expressed as rational polynomials, which correspond to systems expressed as linear, constant-coefficient difference equations (Oppenheim and Schafer, 1989). The crosstalk canceller can be implemented using a network of sample delays, constant gains, and summing

---

[†] A crosstalk canceller that depends on head position will be linear and time varying.

junctions. If the network contains no feedback loops, then it is guaranteed to be realizable, which means that each set of output samples can be computed from the set of input samples and the state of the internal delays. The system will also be stable. The stability and realizability of the network are only issues when the network contains feedback loops. A simple feedback loop is shown in figure 3.18, and it has the following z-transform:

$$H(z) = \frac{1}{1 - A(z)} = 1 + A(z) + A^2(z) + \ldots \quad (3.23)$$

For the system in figure to be realizable, the feedback loop must contain at least one sample delay, otherwise it is impossible to compute the current output. This means that $A(z)$, expressed as a polynomial in z^{-1}, must contain a common factor of z^{-1}. Referring back to the crosstalk cancellation solution in equation 3.19, if the term $1/(1 - ITF_L ITF_R)$ is implemented using a feedback loop, then this will be realizable if the cascade of the two ITFs contains at least one sample of delay. Assuming an ITF can be modeled as a causal filter cascaded with a delay, then the condition for realizability is that the sum of the two interaural time delays be greater than zero:

$$ITD_L + ITD_R > 0 \quad (3.24)$$

The ITD is positive for positive incident angles, and increases monotonically with increasing lateral angle of incidence.

It is easy to see that the realizability constraint of equation 3.24 is met when the listener is facing forward. Figure 3.19 shows the standard listening situation when the listener's head is rotated θ_h degrees right. θ_L and θ_R give the incident angles for ITF_L and ITF_R, respectively. When the listener is facing between the speakers, both incident angles are positive, therefore both interaural time delays are also positive, and the realizability constraint is easily met. When the head is rotated just beyond a speaker, the ITD for that side becomes negative, while the opposite side ITD stays positive, and because of the monotonicity property, the sum of the ITDs stays positive. According to a spherical head model for ITDs, the ITDs become equal and opposite in sign when the head is oriented at ±90 degrees. Thus, the realizability constraint of equation 3.24 is met when $-90 < \theta_h < 90$.

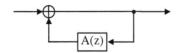

Figure 3.18 Simple feedback loop.

3-D Audio Using Loudspeakers

A necessary condition for stability of the crosstalk canceller is that all poles of the system's z-transform have magnitude less than 1. The region of convergence (ROC) then includes the unit circle, from which it follows that the system impulse response is absolutely summable, and therefore the system is stable in the bounded-input bounded-output (BIBO) sense (Oppenheim and Schafer 1989). Consider the simple feedback loop in figure 3.18 whose z-transform is given in equation 3.23. The geometric series will converge if and only if $|A(z)| < 1$, and therefore the ROC includes the unit circle if and only if $|A(e^{j\omega})| < 1$ for all ω.

Applying this constraint to equation 3.19, the crosstalk canceller will be stable if and only if

$$\left|ITF_L(e^{j\omega})\right|\left|ITF_R(e^{j\omega})\right| < 1, \forall \omega \tag{3.25}$$

The ITF describes head shadowing, and for positive incident angles and low frequencies the ITF is basically a lowpass filter which rolls off with increasing frequency. Furthermore, at low frequencies and small incident angles, the ITF magnitude decreases monotonically with increasing lateral angle of incidence (see figure 3.13 on page 45). For a symmetrical head model, ITFs at negative incident angles are the inverses of the corresponding positive incident ITFs, and are highpass filters. At high frequencies the magnitude of the ITF may be greater than 1, even at positive incident angles, because of notches in the ipsilateral response (see figure 3.8 on page 38). Thus, to ensure a stable crosstalk canceller it may be necessary to either limit the gain of the ITF model, or to use a bandlimited ITF model, the latter being the approach we will take. Considering then only the low-frequency portion of the ITF, we find that the constraint in equation 3.25 is met for frontal head orientations. When the listener is facing between the speakers, both incident angles are positive, and both ITFs are lowpass filters. When the head rotates just beyond a speaker, the ITF for that side

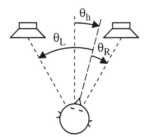

Figure 3.19 Incident angles of speakers for rotated head. θ_L and θ_R are incident angles of left and right speakers, respectively, and θ_h is angle of head. In this example, all three angles are positive.

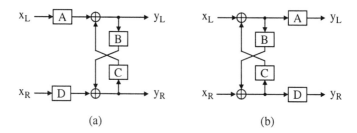

Figure 3.20 Recursive topologies for implementing the 2x2 inverse matrix. The symmetric form of (a) has been used by Iwahara and Mori (1978) to implement crosstalk cancellation filters.

becomes highpass, but the opposite side ITF is still lowpass, and because of the monotonicity property, the product of the ITFs will still have magnitude less than 1. According to a spherical model for ITFs, the ITFs in equation 3.25 become reciprocals when the head rotates to ±90 degrees. Thus, the stability constraint of equation 3.25 is met for low frequencies when $-90 < \theta_h < 90$.

A simpler way to reach this result is to consider the head transfer matrix **H** for a spherical head model. When the head is rotated to ±90 degrees, both speakers fall in the same "cone of confusion," the columns of the matrix **H** become equal, and **H** therefore becomes singular and non-invertible. We expect that a real head model will behave similarly at low frequencies, i.e. that **H** will become singular, or at least ill-conditioned, for head orientations near ±90 degrees.

Note that when the head is rotated beyond ±90 degrees to face the rear, both the realizability and stability constraints can be met if the left and right output channels are swapped. This corresponds exactly to implementing a crosstalk cancellation system using a pair of rear loudspeakers.

3.3.4 Recursive topologies

The straightforward way to implement the 2x2 inverse matrix of equation 3.18 is using four feedforward filters, as shown in figure 3.17. Two recursive filter topologies which can also implement the inverse matrix are shown in figure 3.20. The symmetric form of figure 3.20a has been used by Iwahara and Mori (1978) to implement crosstalk cancellers. These recursive topologies are also commonly used to implement adaptive filters for blind source separation (e.g., see Torkkola, 1996).

The system equations for the topology in figure 3.20a are:

3-D Audio Using Loudspeakers

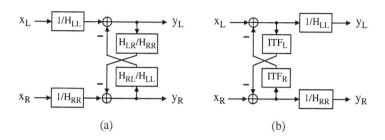

Figure 3.21 Recursive implementations of the asymmetric crosstalk cancellation filter. The symmetric form of (a) has been used by Iwahara and Mori (1978).

$$y_L = Ax_L + Cy_R$$
$$y_R = Dx_R + By_L \quad (3.26)$$

The coefficients in equation 3.26 can be solved to satisfy equation 3.19. The solutions are:

$$A = \frac{1}{H_{LL}}$$
$$B = -\left(\frac{H_{LR}}{H_{RR}}\right)$$
$$C = -\left(\frac{H_{RL}}{H_{LL}}\right) \quad (3.27)$$
$$D = \frac{1}{H_{RR}}$$

The implementation is shown in figure 3.21a. The cross-coupled feedback filters are the HRTF ratios encountered in equation 3.22, and the feedforward filters are the inverse ipsilateral responses.

The system equations for the topology in figure 3.20b are:

$$y_L = A\left(x_L + \frac{C}{D}y_R\right)$$
$$y_R = D\left(x_R + \frac{B}{A}y_L\right) \quad (3.28)$$

The coefficients in equation 3.28 can be solved to satisfy equation 3.19. The solutions are:

$$A = \frac{1}{H_{LL}}$$

$$B = -\left(\frac{H_{LR}}{H_{LL}}\right)$$

$$C = -\left(\frac{H_{RL}}{H_{RR}}\right) \quad (3.29)$$

$$D = \frac{1}{H_{RR}}$$

The implementation is shown in figure 3.21b. The cross-coupled feedback filters are ITFs. Although both implementations are mathematically equivalent, figure 3.21b is far more intuitive. As described earlier, convolving either channel with the appropriate ITF predicts the crosstalk that will reach the contralateral ear. The crosstalk is then cancelled by feeding the negative of this predicted signal into the opposite channel. An important feature of this circuit is that it feeds the cancellation signal back to the opposite channel's input rather than its output, and thus higher-order crosstalks are automatically cancelled. Finally, each channel output is equalized with the corresponding inverse ipsilateral response.

3.3.5 Symmetric solutions

Most of the implementations discussed in the literature assume a symmetric listening situation. Obviously, the symmetric solution is simply a particular case of the general solution, but consideration of symmetry can lead to simplified implementations. When the listening situation is symmetric, we define:

$$H_i = H_{LL} = H_{RR}, H_c = H_{LR} = H_{RL} \quad (3.30)$$

where H_i is the ipsilateral transfer function, and H_c is the contralateral transfer function. Substituing the symmetric variables into equation 3.18, we obtain:

$$\mathbf{H}^{-1} = \begin{bmatrix} H_i & -H_c \\ -H_c & H_i \end{bmatrix} \frac{1}{H_i^2 - H_c^2} \quad (3.31)$$

Dividing by H_i^2, we obtain:

$$\mathbf{H}^{-1} = \begin{bmatrix} 1 & -ITF \\ -ITF & 1 \end{bmatrix} \frac{1/H_i}{1 - ITF^2} \quad (3.32)$$

where

3-D Audio Using Loudspeakers

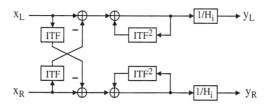

Figure 3.22 Implementation of symmetric crosstalk cancellation filter (Schroeder, 1973).

$$ITF = \frac{H_c}{H_i} \qquad (3.33)$$

is the interaural transfer function for the symmetrical situation. This symmetric formula was described by Schroeder (1973). The corresponding flow diagram is shown in figure 3.22.

Cooper and Bauck (1989) later proposed using a "shuffler" implementation of the crosstalk canceller, which involves forming the sum and difference of the binaural inputs, filtering these signals, and then undoing the sum and difference operation (Blumlein, 1933). The generic shuffler filter circuit is shown in figure . The sum and difference operation is accomplished by the unitary matrix **U** below, called a shuffler matrix:

$$\mathbf{U} = \begin{bmatrix} 1 & 1 \\ 1 & -1 \end{bmatrix} \frac{1}{\sqrt{2}} \qquad (3.34)$$

Columns of the matrix **U** are eigenvectors of the symmetric 2x2 matrix, and therefore the shuffler matrix **U** diagonalizes the symmetric matrix $\mathbf{H}^{-1}$ via a similarity transformation (e.g., see Horn and Johnson, 1985):

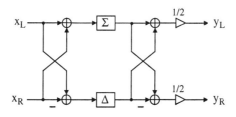

Figure 3.23 Shuffler filter structure. This has been used for implementing crosstalk cancellers by Cooper and Bauck (1989).

$$\mathbf{H}^{-1} = \mathbf{U}^{-1} \begin{bmatrix} \dfrac{1}{H_i + H_c} & 0 \\ 0 & \dfrac{1}{H_i - H_c} \end{bmatrix} \mathbf{U} \quad (3.35)$$

Thus, the crosstalk canceller is implemented with shuffler filters Σ and Δ that are the inverses of the sum and difference of the ipsilateral and contralateral responses (Cooper and Bauck, 1989):

$$\begin{aligned} \Sigma &= 1/(H_i + H_c) \\ \Delta &= 1/(H_i - H_c) \end{aligned} \quad (3.36)$$

The shuffler topology is shown in figure 3.23. The $1/\sqrt{2}$ normalizing gains have been commuted to a single gain of 1/2 for each channel. Note that $\mathbf{U} = \mathbf{U}^{-1}$, so the same sum and difference operation appears on both sides of the Σ and Δ filters.

A further simplification to equation 3.35 can be made by factoring out $1/H_i$, which yields:

$$\mathbf{H}^{-1} = \mathbf{U}^{-1} \begin{bmatrix} \dfrac{1}{1 + ITF} & 0 \\ 0 & \dfrac{1}{1 - ITF} \end{bmatrix} \mathbf{U} \dfrac{1}{H_i} \quad (3.37)$$

This formulation has been suggested by Jot (1992) and subsequently by Gardner (Casey et al., 1995). The ITF can be modeled as an interaural time delay cascaded with a lowpass head-shadowing filter. The shuffler filters are then seen to be simple comb filters with lowpass filters in the feedback loops, with the following transfer functions:

$$\begin{aligned} \Sigma &= \dfrac{1}{1 + ITF} \\ \Delta &= \dfrac{1}{1 - ITF} \end{aligned} \quad (3.38)$$

In practice, the inverse ipsilateral response in equation 3.37 can be commuted back to the binaural synthesis stage by using synthesis HRTFs which are free-field equalized with respect to the loudspeaker direction, as suggested by Jot[†].

[†]Jean-Marc Jot, personal communication, 1996.

3-D Audio Using Loudspeakers

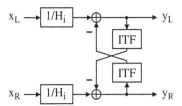

Figure 3.24 Symmetric recursive structure (Iwahara and Mori, 1978). The inverse ipsilateral filters can be associated with the inputs or outputs of the system.

The recursive structures in figure 3.21 can of course be used for the symmetric solution, and this has been described by Iwahara and Mori (1978). When the system is symmetric, both feedback filters become the ITF, and the inverse ipsilateral filter can be associated with either the inputs or outputs of the system. In a symmetric implementation, it always makes sense to commute the inverse ipsilateral filter to the binaural synthesis filters by using free-field equalized HRTFs. Figure 3.24 shows this symmetric recursive structure.

For the symmetric case, the condition for crosstalk cancellation analogous to the constraint in equation 3.22 is that the ITF of the listener equal the ITF of the crosstalk cancellation head model.

3.4 BANDLIMITED IMPLEMENTATIONS

3.4.1 Bandlimiting the crosstalk canceller

The general solution to the crosstalk canceller given in equation 3.17 can be bandlimited so that the crosstalk cancellation only functions for low frequencies. One method is given below:

$$\mathbf{C} = H_{LP}\mathbf{H}^{-1} + H_{HP}\begin{bmatrix} 1 & 0 \\ 0 & 1 \end{bmatrix} \qquad (3.39)$$

where H_{LP} and H_{HP} are lowpass and highpass filters, respectively, with complementary magnitude responses. Thus, at low frequencies $\mathbf{C}$ is equal to $\mathbf{H}^{-1}$ and at high frequencies $\mathbf{C}$ is equal to the identity matrix. This means that crosstalk cancellation and ipsilateral equalization occur at low frequencies, and at high frequencies the binaural signals are passed unchanged to the loudspeakers. Another method is as follows:

$$\mathbf{C} = \begin{bmatrix} H_{LL} & H_{LP}H_{RL} \\ H_{LP}H_{LR} & H_{RR} \end{bmatrix}^{-1} \quad (3.40)$$

Here the cross-terms of the head transfer matrix are lowpass filtered prior to inversion. This is essentially the approach proposed by Cooper and Bauck (1990). Applying a lowpass filter to the contralateral terms has the effect of replacing each ITF term in equation 3.19 with a lowpass filtered ITF. This yields filters which are easy to implement; we have already seen that it is easy to create causal filters which closely approximate lowpass filtered ITFs.

Using the bandlimited form of equation 3.40, at low frequencies $\mathbf{C}$ is equal to $\mathbf{H}^{-1}$, but now at high frequencies $\mathbf{C}$ continues to implement the ipsilateral equalization:

$$\mathbf{C}_{f > f_c} = \begin{bmatrix} 1/H_{LL} & 0 \\ 0 & 1/H_{RR} \end{bmatrix} \quad (3.41)$$

This formulation is important when we attempt to build systems that observe the power panning property. Using equation 3.40, when the sound is panned to the location of a speaker, the response to that speaker will be flat, as desired. Unfortunately, the other speaker will be emitting power at high frequencies, because the crosstalk canceller is not implementing the inverse matrix. We will later describe a method that re-establishes the power panning property at high frequencies.

The symmetric crosstalk canceller is bandlimited in exactly the same way as the asymmetric filter. Following the preferred method of equation 3.40, the bandlimited symmetric crosstalk canceller is:

$$\mathbf{C} = \begin{bmatrix} H_i & H_{LP}H_c \\ H_{LP}H_c & H_i \end{bmatrix}^{-1} \quad (3.42)$$

This leads to replacing the ITF in equation 3.32 with a lowpass filtered ITF.

3.4.2 Recursive asymmetric bandlimited implementation

In this section we will discuss a recursive filter structure that implements an asymmetric bandlimited crosstalk canceller. The structure is composed of causal bandlimited ITFs that model head shadowing, integer and fractional delays, and minimum-phase equalization filters. The structure is created by using the lowpass ITF model of equation 3.7 in the recursive topology of figure 3.21. This is shown in figure 3.25,

3-D Audio Using Loudspeakers

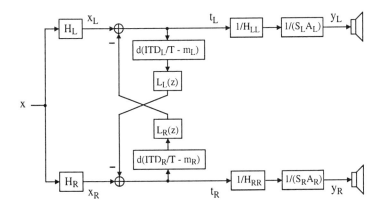

Figure 3.25 Using bandlimited ITFs in a recursive topology. $L_L(z)$ and $L_R(z)$ are causal head shadowing filters with modeling delays of m_L and m_R, respectively. $d(p)$ is a delay of p samples.

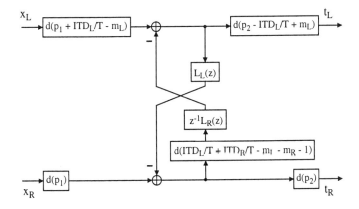

Figure 3.26 Coalescing the total loop delay in figure 3.25 to a single delay. p_1 and p_2 are integer modeling delays.

along with a single source binaural synthesizer and the speaker and air propagation inverse filters.

We introduce the notation $d(p)$ as implementing an integer or fractional delay of p samples. The delays m_L and m_R are modeling delays inherent in the head shadowing filters $L_L(z)$ and $L_R(z)$, respectively. The structure of figure 3.25 is only realizable when both feedback delays are greater than 1, which is much more restrictive than the condition that the sum of the ITDs be positive (equation 3.24). To allow one of the ITDs to become negative, we need to coalesce the total loop delay into a single delay.

This is easily done and the result is shown in figure 3.26. It is necessary to add the integer modeling delays p_1 and p_2 such that:

$$p_1 + \frac{ITD_R}{T} - m_R > 0$$
$$p_2 - \frac{ITD_R}{T} + m_R > 0$$
(3.43)

A single sample delay remains cascaded with $L_L(z)$. The realizability constraint is then:

$$\frac{ITD_L}{T} + \frac{ITD_R}{T} - m_L - m_R - 1 \geq 0 \qquad (3.44)$$

This constraint accounts for the single sample delay remaining in the loop and the modeling delays inherent in the lowpass head shadowing filters.

We now turn our attention to the output side of figure 3.25, namely the ipsilateral equalization filters, and the speaker and air propagation inverse filters. It is important to note that the ipsilateral equalization filters not only provide high-frequency spectral equalization, but also compensate for the asymmetric path lengths to the ears when the head is rotated. We would like to separately parameterize these asymmetric delays but we do not have a parameterized model for an HRTF or its inverse. However, we can use Jot's model for ratios of HRTFs which is accurate at low frequencies (equation 3.6). In order to convert the ipsilateral equalization filters to ratios, we can use free-field equalized synthesis HRTFs, and then the ipsilateral equalization filters become referenced to the free-field direction.

It is most convenient to reference the synthesis HRTFs with respect to the default loudspeaker direction, which we notate as θ_s, i.e., $\theta_s = 30$ degrees for the conventional listening geometry. Therefore, we use H_X/H_{θ_s} for the synthesis filter in channel $X \in \{L, R\}$ and the corresponding ipsilateral equalization filter becomes H_{θ_s}/H_{XX}, where H_{θ_s} is the ipsilateral HRTF for the speaker incidence angle θ_s. Note that H_{θ_s} depends only on the listening geometry, not on the rotation angle of the head. When the head is not rotated, $H_{XX} = H_{\theta_s}$, and the ipsilateral equalization filter will be flat. With this choice of a free-field reference, we now apply the model of equation 3.6:

$$\frac{H_{\theta_s}(e^{j\omega})}{H_{XX}(e^{j\omega})} \cong \text{minp}\left(\frac{H_{\theta_s}(e^{j\omega})}{H_{XX}(e^{j\omega})}\right) e^{-j\omega b_X} \quad (3.45)$$

where b_X is the delay in samples for ear X relative to the unrotated head position.

Included for analytical rigor, the speaker inverse filters $1/S_X$ are often ignored in practice. A robust implementation could include the speaker inverse filters to compensate for asymmetries in the speaker responses. Even with perfectly matched loudspeakers, non-uniform directivity patterns can cause asymmetrical responses for off-axis listeners.

The air propagation inverse filters $1/A_X$ are very important, because they compensate for unequal path lengths from the speakers to the center of the head. A simple model for the air propagation consisting of a delay and an attentuation is accurate:

$$A_X(e^{j\omega}) = k_X e^{-j\omega a_X} \quad (3.46)$$

The combined ipsilateral equalization and air propagation inverse filter for channel X is now:

$$\frac{H_{\theta_s}(e^{j\omega})}{H_{XX}(e^{j\omega})} \cdot \frac{1}{A_X(e^{j\omega})} \cong \frac{1}{k_X}\text{minp}\left(\frac{H_{\theta_s}(e^{j\omega})}{H_{XX}(e^{j\omega})}\right) e^{-j\omega(b_X - a_X)} \quad (3.47)$$

One final simplification is to lump all of the variable output delay into the left channel. This is done by associating a variable delay of $a_L - b_L$ with both channels. This means that head motions that change the difference in path lengths from the speakers to the ears will induce a slight but unnoticable pitch shift in both output channels. Delaying and attenuating an output channel to compensate for a displaced listener is a well known technique (Cohen, 1982; Cooper and Bauck, 1990). The final and complete implementation is shown in figure 3.27. The modeling delay p_2 must now be increased so that:

$$p_2 - \frac{ITD_R}{T} + m_R + b_L - a_L - b_R + a_R \geq 0 \quad (3.48)$$

The structure in figure 3.27 compensates for any translation of the head center and any head orientation with respect to the speakers, provided the realizablility and stability constraints are met. The crosstalk canceller is implemented with two lowpass head shadowing filters, three fractional delays, two fixed delays, and two minimum-phase ipsilateral equalization filters. An accurate implementation of each component in the structure will yield excellent performance because we have used very conservative simplifying assumptions. Consequently, the structure is capable of rendering an individualized crosstalk cancellation head model. The real significance of this structure is that we may replace each component with a much simplified implementation to arrive at a computationally efficient structure that has adequate performance. The lowpass head shadowing filters may be implemented using low-order IIR filters. The fractional delay lines may be implemented using low-order FIR interpolators. The minimum phase ipsilateral equalization filters may be omitted entirely, but this will degrade the crosstalk cancellation for rotated head orientations. These filters are performing important phase compensation at low frequencies.

3.4.3 Parameter dependency on head position

In the circuit of figure 3.27, the filter functions H_X, H_{XX}, and $L_X(z)$, as well as the delay parameters ITD_X, a_X, and b_X, and the gains k_X, are all dependent on the current position and orientation of the head. For simplicity, we assume the head is vertically oriented in the horizontal plane of the speakers so that the head position and orientation are fully specified by the (x, y) position of the head center and the head rotation angle. Note, however, that the specification of the crosstalk canceller generalizes to arbitrary head orientations with respect to the speakers.

The filter functions, delay parameters, and gains should be stored in pre-computed tables. When the listener's head moves, the current position and orientation of the listener's head, as detected by the head tracker, are used to access the stored parameters and update the crosstalk canceller. All of the parameters must be updated in a way that avoids spurious transients.

Two approaches to smoothly updating filter functions are *interpolation* and *commutation*, using the terminology of Jot et al. (1995). Interpolation between adjacent filter functions is needed when the pre-computed filters are sparsely sampled; interpolation can also be used to smooth the update transitions, by slowly interpolating the filter coefficients from the old values to the new values. Commutation refers to switching from the old to the new filter function with a short crossfade to eliminate transients; this crossfade requires that both the old and new filters run concurrently during the transition period. We have generally adopted the commutation approach to filter updates, in part because the KEMAR HRTF data are densely sampled and therefore interpolation between stored filter functions is not necessary. However, either approach, or a combination of the two, may be used.

The $L_X(z)$ head-shadowing filters and the ipsilateral equalization filters both depend on the position and orientation of the listener's head. Each binaural synthesis filter also depends on the position and orientation of the listener's head, as well as the target location of the source. The dependency on the listener is encountered in all head-tracked spatial auditory displays; the synthesis location is chosen relative to the cur-

3-D Audio Using Loudspeakers

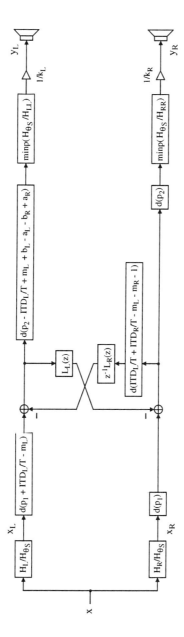

Figure 3.27 Recursive implementation of asymmetric bandlimited crosstalk canceller. This structure compensates for front-back head motion, lateral head motion, and head rotations. The crosstalk canceller is implemented with two lowpass head shadowing filters, three fractional and two fixed delays, and two minimum-phase ipsilateral equalization filters.

rent head orientation so as to synthesize an externally stationary sound during head motion.

Provided the head tracking updates are reasonably frequent, for instance, 60 updates/sec, the delay parameters will change only slightly with each update and will not require interpolation. The fractional delay lines are best implemented using low-order FIR interpolators (Laakso et al., 1996). Third-order Lagrangian interpolation is preferable to first-order linear interpolation because the latter method produces audible high-frequency modulation during delay changes.

The ITD_X parameter can be calculated from a spherical head model, as shown in equation 3.1. Alternatively, the ITD can be calculated from pre-computed ITFs, by performing a linear regression on the interaural excess phase (Jot et al., 1995), as discussed in section 3.2.5. A comparison of the two methods is shown in figure 3.3.

The parameter b_X is a function of head angle, the constant parameter θ_S (the absolute angle of the speakers with respect to the listener when in the ideal listening location), and the constant parameter f_s (the sampling rate). The parameter b_X represents the delay (in samples) of sound from speaker X reaching the ipsilateral ear, relative to the delay when the head is in the ideal (unrotated) listening location. Like ITD_X, b_X may be calculated from a spherical head model; the result is a trigonometric function:

$$b_R(\theta_H) = -\frac{Df_s}{2c}(s(\theta_H - \theta_S) + s(\theta_S)) \qquad (3.49)$$

where θ_H is the rotation angle of the head, such that $\theta_H = 0$ when the head is facing forward, D is the diameter of the head in meters, c is the speed of sound in m/sec, and the function $s(\theta)$ is defined as:

$$s(\theta) = \begin{cases} \sin\theta, & \theta < 0 \\ \theta, & \theta > 0 \end{cases} \qquad (3.50)$$

The function $b_L(\theta)$ is defined as $b_R(-\theta)$. An alternative to using the spherical head model is to compute the b_X parameter by performing linear regression on the excess phase part of the ratio H_{θ_s}/H_{XX}. This is completely analogous to the technique used to determine the ITD from a ratio of two HRTFs. Figure 3.28 shows both methods of computing b_R for head azimuths from -90 to +90 degrees, with $\theta_S = 30$ degrees, $f_s = 44100$: the solid line is the geometrical model of equation 3.49; the

3-D Audio Using Loudspeakers

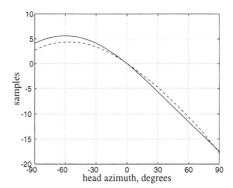

Figure 3.28 Plot of b_R parameter with $\theta_S = 30$ degrees, $f_s = 44100$: the solid line is the geometrical model; the dashed line is the result of performing linear regression on the excess phase part of the ratio of the appropriate HRTFs

dashed line is the result of performing linear regression on the excess phase part of the ratio of the appropriate HRTFs.

The parameters a_X and k_X are functions of the distances d_L and d_R between the center of the head and the left and right speakers, respectively. These distances are provided along with the head rotation angle by the head tracker. According to equation 3.46, a_X represents the air propagation delay in samples between speaker X and the center of the head, and k_X is the corresponding attenuation in sound pressure due to the air propagation. Without loss of generality, we can normalize these parameters with respect to the ideal listening location such that $a_X = 0$ and $k_X = 1$ when the listener is ideally situated. The equations for a_X and k_X are then:

$$a_X = \frac{f_s(d_X - d)}{c}$$
$$k_X = \frac{d}{d_X}$$
(3.51)

where d_X is the distance from the center of the head to speaker X, expressed in meters, and d is the distance from the center of the head to the speakers when the listener is ideally situated, also expressed in meters.

3.4.4 Feedforward asymmetric bandlimited implementation

A straightforward method of implementing the bandlimited crosstalk cancellation filter is to use the feedforward form of figure 3.17 on page 49 and to add lowpass filters to the cross terms (Cooper and Bauck, 1990). This is shown in figure 3.29. In this

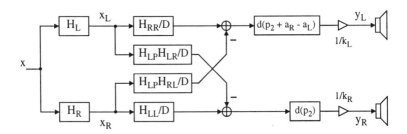

Figure 3.29 Feedforward implementation of asymmetric bandlimited crosstalk canceller. The four feedforward filters are implemented with pre-computed FIR filters. H_{LP} is a lowpass filter. Cooper and Bauck (1990) proposed a similar bandlimited symmetric structure.

implementation, the inverse determinant filter has been incorporated into each of the feedforward filters in order to reduce the number of individual filters required. Each of the feedforward filters can be implemented using an FIR filter. A set of feedforward filters is pre-computed for a specific listening geometry, which may be parameterized in terms of the head rotation angle and the angular spread of the speakers with respect to the head. Asymmetrical path lengths to the head are compensated for using the variable delay line and variable output gains described earlier.

The advantage of this approach is that it is trivial to interpolate between different sets of FIR filters as the head moves. The filters are also relatively easy to design. The inverse determinant filter can be designed using the DFT method described in Appendix A. At a 32 kHz sampling rate, an FIR length of 128 points (4 msec) gives excellent performance. This length of FIR filter can be most efficiently computed using a DFT convolution technique. Per channel, one forward and one inverse DFT needs to be computed, along with two spectral products and one spectral addition. This is only slightly more expensive than a single FIR filter.

3.4.5 Symmetric bandlimited implementations

The symmetric crosstalk canceller can be bandlimited following either of the two methods shown in equation 3.39 and equation 3.40. Following the preferred method of equation 3.40, we arrive at:

$$\mathbf{C} = \begin{bmatrix} H_i & H_{LP}H_c \\ H_{LP}H_c & H_i \end{bmatrix}^{-1} \tag{3.52}$$

which, as in the asymmetric case, leads to replacing the ITF in equation 3.32 with a lowpass filtered ITF. Following equation 3.37 and equation 3.38, this leads to a shuffler implementation with the following shuffler filters:

3-D Audio Using Loudspeakers

$$\Sigma = \frac{1}{1 + H_{LP}ITF}$$
$$\Delta = \frac{1}{1 - H_{LP}ITF}$$
(3.53)

This method was presented by Gardner (Casey et al., 1995). A complete implementation is obtained following the same strategy used in the previous section; the lowpass ITF model of equation 3.7 is used, the speaker inverse filters are omitted, and the air propagation inverse filters are replaced with a variable delay and gain. This leads to the shuffler implementation shown in figure 3.30. The constraint for realizability is:

$$\frac{ITD}{T} - m - 1 \geq 0 \qquad (3.54)$$

where m is the modeling delay inherent in the lowpass head shadowing filter $L(z)$. The structure can compensate for front-back and lateral head motions, but not head rotations. The crosstalk canceller requires only three fractional delays and two lowpass filters. An even more efficient implementation is shown in figure 3.31, which is mathematically equivalent to figure 3.30, but does not require the shuffler sum and difference structures. This is derived from the symmetric recursive structure shown in figure 3.24 (Iwahara and Mori, 1978).

An alternative to using the lowpass ITF model within the shuffler comb filters is to directly calculate the shuffler filters from:

$$\Sigma = \frac{H_i}{H_i + H_{LP}H_c}$$
$$\Delta = \frac{H_i}{H_i - H_{LP}H_c}$$
(3.55)

This equation is obtained from equation 3.53 by multiplying both the numerator and denominator by H_i. The calculation of the filter responses is easily accomplished in the frequency domain; the corresponding time responses can then be modeled using IIR filters by applying Prony's method. Excellent results can be obtained using 8th-order filters. The resulting shuffler filters are efficient and accurate, but lack a separate ITD parameter. This method yields essentially the same result as using the comb filter approach where the modeling delay m of the ITF is maximized ($m = ITD - 1$).

An extremely efficient head model for bandlimited crosstalk cancellers has been suggested by Griesinger[†] and described by Gardner (1995c). The idea is to model the ITF as a delay cascaded with a one-pole lowpass filter and an attenuating gain:

$$ITF(e^{j\omega}) = gH_{LP}(e^{j\omega})e^{-j\omega(ITD/T-m)}$$

$$H_{LP}(z) = \frac{1-a}{1-az^{-1}} \tag{3.56}$$

The lowpass filter is DC-normalized. The coefficient a determines the filter cutoff, which is typically set to 1-2 kHz; the coefficient g determines the attenuation, which is typically set to 1-3 dB. In practice, the ITD parameter may be determined from a geometrical model, which accurately models the high-frequency (f > 1500 Hz) ITD. Below 1500 Hz, the true ITD is larger by roughly 3/2, as explained in section 3.2.5. Interestingly, the one-pole lowpass filter also has a non-linear phase response with increasing phase delay at lower frequencies. The parameter m may be adjusted to approximately match the total phase delay of the head shadowing model to the desired ideal (we typically set m to 2 samples = 0.05 msec at 44.1 kHz). This ITF model is trivially substituted into the circuits of figure 3.30 and figure 3.31, or even the asymmetric circuit of figure 3.27. The resulting structures are extremely efficient, and yet are effective at cancelling crosstalk at frequencies up to 6 kHz.

3.4.6 High-frequency power transfer model

Using the form of the crosstalk canceller given in equation 3.14, which includes the speaker and air propagation inverse filters, the bandlimited crosstalk canceller is:

$$\mathbf{C} = \mathbf{S}^{-1}\begin{bmatrix} H_{LL} & H_{RL}H_{LPF} \\ H_{LR}H_{LPF} & H_{RR} \end{bmatrix}^{-1} \tag{3.57}$$

At high frequencies, $\mathbf{C}$ becomes:

$$\mathbf{C}_{f>f_c} = \mathbf{S}^{-1}\begin{bmatrix} 1/H_{LL} & 0 \\ 0 & 1/H_{RR} \end{bmatrix} \tag{3.58}$$

As previously described, this implements ipsilateral equalization. The speaker signals for a given source x are:

$$\mathbf{y} = \mathbf{C}\mathbf{h}x \tag{3.59}$$

[†]David Griesinger, personal communication, 1995.

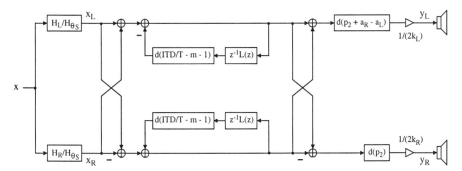

Figure 3.30 Symmetric bandlimited implementation using shuffler topology and parameterized head shadowing model.

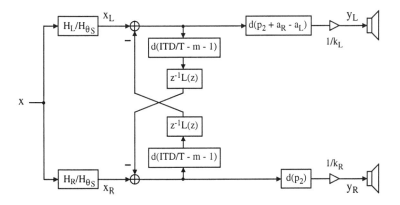

Figure 3.31 Symmetric bandlimited implementation using recursive topology and parameterized head shadowing model. This is mathematically equivalent to the implementation in figure 3.30.

where **h** are the HRTFs for the source position. Thus, at high frequencies:

$$\mathbf{y}_{f>f_c} = \mathbf{S}^{-1} \begin{bmatrix} H_L/H_{LL} \\ H_R/H_{RR} \end{bmatrix} x \tag{3.60}$$

The system responses prior to the equalization $\mathbf{S}^{-1}$ correspond to free-field equalized HRTFs. Essentially, the speakers are emitting binaural signals. When the sound is panned to the location of either speaker, the response to that speaker will be flat, because of the ipsilateral equalization, and this agrees with the power panning property. However, the other speaker will be emitting the free-field equalized contralateral binaural response, which violates the power panning property. If the crosstalk cancellation extended to high frequencies, the contralateral response would be internally cancelled and would not appear at the loudspeaker. Despite this nice property, it is not a good idea to extend crosstalk cancellation to high frequencies. An alternative approach, based on power transfer to the ears, can be used to optimize the presentation of high frequencies and also satisfy the power-panning property.

We assume that high-frequency signals from the two speakers add incoherently at the ears. We model the high-frequency power transfer from the speakers to the ears as a 2x2 matrix of power gains derived from the KEMAR HRTFs. An implicit assumption is that the KEMAR head shadowing is similar to the head shadowing of a typical human. The power transfer matrix is inverted to calculate what powers to send to the speakers in order to get the proper powers at the ears. Often it is not possible to synthesize the proper powers at the ears, for example when synthesizing a right source which is more lateral than the right loudspeaker. In this case the desired ILD is greater than that achieved by sending the signal only to the right loudspeaker. Any power emitted by the left loudspeaker will decrease the final ILD at the ears. In such cases where there is no exact solution, we send the signal to one speaker, and scale its power so that the total power transfer to the two ears equals the total power in the synthesis HRTFs. Except for this caveat, the power formulation is entirely analogous to the usual transmission path inversion we encounter in crosstalk cancellers.

The high-frequency power to each speaker is controlled by associating a multiplicative gain with each output channel. Because the crosstalk canceller is diagonal at high frequencies, the scaling gains can be commuted to the synthesis HRTFs. The scaling gains g_L and g_R are inserted into the previous equation as follows:

$$\mathbf{y} = \mathbf{S}^{-1} \begin{bmatrix} g_L H_L / H_{LL} \\ g_R H_R / H_{RR} \end{bmatrix} x \quad (3.61)$$

The signals at the ears are given by:

$$\mathbf{e} = \mathbf{A}\mathbf{y} = \mathbf{H}\mathbf{S}\mathbf{y} \quad (3.62)$$

where we model the acoustical transfer matrix $\mathbf{H}$ using KEMAR HRTFs. Combining the previous two equations, we arrive at:

$$\begin{bmatrix} e_L \\ e_R \end{bmatrix} = \begin{bmatrix} H_{LL} & H_{RL} \\ H_{LR} & H_{RR} \end{bmatrix} \begin{bmatrix} g_L H_L / H_{LL} \\ g_R H_R / H_{RR} \end{bmatrix} x \quad (3.63)$$

We now convert this equation to an equivalent expression in terms of power transfer. The simplest approach is to model the input signal x as stationary white noise and to assume that the transfer functions to the two ears are uncorrelated. We rewrite equation 3.63 in terms of signal variance by replacing the transfer functions with their corrsponding energies (e.g., Papoulis, 1991):

$$\begin{bmatrix} \sigma_{e_L}^2 \\ \sigma_{e_R}^2 \end{bmatrix} = \begin{bmatrix} E_{H_{LL}} & E_{H_{RL}} \\ E_{H_{LR}} & E_{H_{RR}} \end{bmatrix} \begin{bmatrix} g_L^2 E_{H_L}/E_{H_{LL}} \\ g_R^2 E_{H_R}/E_{H_{RR}} \end{bmatrix} \sigma_x^2 \quad (3.64)$$

where the energy of a discrete-time signal $h[i]$, with corresponding DFT $H[k]$, is given by

$$E_h = \sum_{i=0}^{N-1} h^2[i] = \frac{1}{N} \sum_{k=0}^{N-1} |H[k]|^2 \quad (3.65)$$

The power transfer to the ears is then

$$\begin{bmatrix} \sigma_{e_L}^2/\sigma_x^2 \\ \sigma_{e_R}^2/\sigma_x^2 \end{bmatrix} = \begin{bmatrix} E_{H_{LL}} & E_{H_{RL}} \\ E_{H_{LR}} & E_{H_{RR}} \end{bmatrix} \begin{bmatrix} g_L^2 E_{H_L}/E_{H_{LL}} \\ g_R^2 E_{H_R}/E_{H_{RR}} \end{bmatrix} \quad (3.66)$$

We replace the actual power transfer to the ears with the desired power transfer corresponding to the synthesis HRTFs and solve for the scaling gains:

$$\begin{bmatrix} E_{H_L} \\ E_{H_R} \end{bmatrix} = \begin{bmatrix} E_{H_{LL}} & E_{H_{RL}} \\ E_{H_{LR}} & E_{H_{RR}} \end{bmatrix} \begin{bmatrix} g_l^2 E_{H_L}/E_{H_{LL}} \\ g_R^2 E_{H_R}/E_{H_{RR}} \end{bmatrix} \quad (3.67)$$

$$\begin{bmatrix} g_L^2 \\ g_R^2 \end{bmatrix} = \begin{bmatrix} E_{H_{LL}}/E_{H_L} & 0 \\ 0 & E_{H_{RR}}/E_{H_R} \end{bmatrix} \begin{bmatrix} E_{H_{LL}} & E_{H_{RL}} \\ E_{H_{LR}} & E_{H_{RR}} \end{bmatrix}^{-1} \begin{bmatrix} E_{H_L} \\ E_{H_R} \end{bmatrix} \quad (3.68)$$

This equation is the crosstalk canceller expressed in terms of broadband power transfer. If either row of the right hand side of the above equation is negative, then a real solution is not attainable. In this case, we set the gain corresponding to the negative row equal to 0, and set the other gain term so that the total power at the ears is equal to the total desired power. The expression relating total desired power and total power follows directly from equation 3.67 by adding the two rows:

$$E_{H_L} + E_{H_R} = g_L^2 \frac{E_{H_L}}{E_{H_{LL}}}(E_{H_{LL}} + E_{H_{LR}}) + g_R^2 \frac{E_{H_R}}{E_{H_{RR}}}(E_{H_{RL}} + E_{H_{RR}}) \quad (3.69)$$

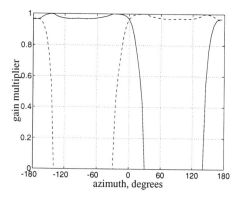

Figure 3.32 High-frequency (> 6 kHz) scaling gains applied to synthesis HRTFs for the standard symmetric listening geometry. Solid line: g_L, dashed line: g_R. For the symmetric listening situation, the scaling gains have the effect of shutting off the contralateral speaker when synthesizing lateral source locations.

This expression is solved for one gain when the other gain is set to 0. Because all energies are non-negative, a real solution is guaranteed.

This theory can be put into practice by creating a set of HRTFs that have the high-frequency response scaled according to the above equations. The listening geometry specifies the head transfer matrix **H**, which is converted to energies by highpass filtering and then applying equation 3.65. For each synthesis location, we read the corresponding HRTFs and separate into low and high-frequency components using zero-phase FIR filters with complementary magnitude responses. The high-frequency components are converted to energies and inserted into equation 3.68 along with the acoustic transfer energies. After solving for the gains, the high-frequency HRTF components are scaled and added to the low-frequency HRTF components. This creates a set of HRTFs with high frequencies adjusted for a particular listening geometry. Scaling the high-frequency components of the synthesis HRTFs in this method corresponds exactly to applying a high-frequency shelving filter to each synthesis HRTF. Efficient implementations are discussed in the next section.

For the standard symmetric listening situation (speakers at ±30°, listener's head rotated 0°), the scaling gains for horizontal sources are shown in figure 3.32. When the sound is panned to the location of a loudspeaker, there is an exact solution which simply sends all power to that speaker and shuts off the other speaker. Thus, application of this theory creates a system that has the power panning property. For sources beyond 30°, there isn't an exact solution, the contralateral gain is 0, and the ipsilateral gain is chosen to conserve total energy.

Figure 3.33a shows the high-frequency power transmitted to the left and right ears for source azimuths from 0 to 180 degrees. The solid and dashed lines show the ipsilateral and contralateral power, respectively, for binaural listening (i.e. these are the

3-D Audio Using Loudspeakers

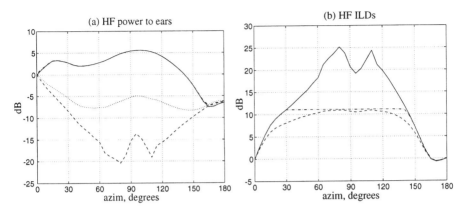

Figure 3.33 (a) High-frequency (> 6 kHz) power transfer to the two ears for horizontal sources, relative to a source at 0 degrees. Solid and dashed lines: ipsilateral and contralateral powers, respectively, of synthesis HRTFs. Dash-dot and dotted lines: ipsilateral and contralateral powers, respectively, when listening to synthesis HRTFs using loudspeakers in standard geometry. Crosstalk negligibly affects the ipsilateral power, but greatly affects the contralateral power. (b) High-frequency (> 6 kHz) interaural level differences (ILDs) for horizontal sources. Solid line: ILD of synthesis HRTFs, dashed line: ILD resulting from presenting binaural signals over loudspeakers, dash-dot line: latter with power compensation applied. The power compensation has the effect of increasing ILD up to 30 degrees azimuth, where the ILD is limited to 11 dB by acoustical crosstalk.

powers in the synthesis HRTFs, normalized to 0 degrees incidence). The ILD is simply the difference between these two lines. Clearly visible is the decrease in high-frequency power for rear sources. The dotted line shows the contralateral power when head crosstalk is factored in, assuming speakers at ±30 degrees, and without any power compensation described above. The dash-dot line (barely visible behind the solid line) is the ipsilateral response when crosstalk is factored in. It is clear that crosstalk greatly increases power to the contralateral ear and thus limits the maximum ILD to be 11dB, which is the ILD for a 30 degree source.

Figure 3.33b shows the high-frequency ILDs that occur in various situations. The solid line is the ILD of the synthesis HRTFs. The dashed line is the ILD at the ears when binaural signals are presented over loudspeakers and crosstalk occurs. The dash-dot line is the ILD that occurs when we employ the power model described above. As expected the proper ILDs are generated up to 30 degrees, after which the ILD remains at the maximum value (contralateral gain = 0).

As indicated in figure 3.33b, implementing the high-frequency power model when the listener's head is not rotated only achieves modest improvements over using the unmodified HRTFs of equation 3.60. However, the high-frequency gain modification is critically important when the listener's head is rotated, otherwise the low and high-frequency components will be synthesized at different locations, the low frequencies relative to the head, and the high frequencies relative to the speakers. Application of the high-frequency power model also has the nice theoretical result that the power panning property holds for all frequencies.

3.4.7 High-frequency implementation

The last section described a method for computing a set of HRTFs whose high frequencies are scaled for a particular listening geometry. This would require that a separate set of synthesis HRTFs be used for each orientation of the head with respect to the speakers. It is more sensible to implement separately a high-frequency shelving filter that operates on each channel of each binaural signal. The gains of each shelving filter are dependent on the listening geometry and the source location, and can be looked up in pre-computed tables. Figure 3.34 shows a pair of shelving filters applied to the binaural signal for a source. It is very important that the two shelving filters have the same low-frequency phase and magnitude response independent of the high-frequency gains, otherwise the shelving filters will induce unwanted interaural differences.

The shelving filter implementation suggested in the last section is shown in figure 3.35a where H_{LP} and H_{HP} are lowpass and highpass filters, respectively. When H_{LP} and H_{HP} have complementary responses, $H_{LP}(z) = 1 - H_{HP}(z)$, and this enables the simplified form of figure 3.35b. Unfortunately, it is not possible to use a low-order IIR lowpass filter for H_{LP} because the low-frequency phase response of the shelving filter will depend on the high-frequency gain. We must therefore use a linear-phase FIR filter for H_{LP}. This will add considerable computation to our implementation. Fortunately, we only need one lowpass filter for each summed binaural channel to implement independent shelving filters for any number of summed sources. From figure 3.35b:

$$\hat{x}_i = g_i(1 - H_{LP})x_i + H_{LP}x_i$$
$$\hat{x}_i = g_i x_i - H_{LP}x_i(1 - g_i)$$
(3.70)

The sum over all sources is then:

$$\sum_i \hat{x}_i = \sum_i g_i x_i - H_{LP}\sum_i x_i(1 - g_i)$$
(3.71)

This leads to the implementation shown in figure 3.36. The figure shows the left channel processing, where x_{Li} is the left channel binaural signal for source i, g_{Li} is the left channel high-frequency scaling gain for source i, $\hat{x}_{Li}$ is the high-frequency adjusted left channel binaural signal, and m is the linear phase delay of H_{LP}. The same circuit is used for the right channel.

3-D Audio Using Loudspeakers

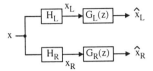

Figure 3.34 High-frequency shelving filters $G_L(z)$ and $G_R(z)$ applied to output of binaural synthesizer.

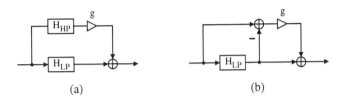

Figure 3.35 High-frequency shelving filter implementations.

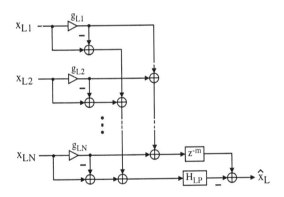

Figure 3.36 High-frequency shelving filters applied to the left channel of all N sources. Only a single lowpass filter is required. H_{LP} is a linear-phase lowpass filter with a phase delay of m samples.

3.4.8 Complete head-tracked system

A complete head-tracked 3-D audio loudspeaker system is created by combining a multiple source binaural synthesizer with a crosstalk canceller and a head tracker; such a system is depicted in figure 3.37. At the left are the N input sounds x_i whose spatial positions are to be separately synthesized. Each sound is filtered with an appropriate HRTF pair (H_{iL}, H_{iR}) to encode directional cues. The equalization of the

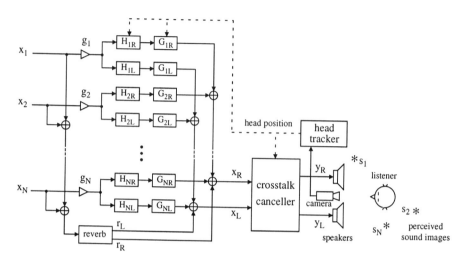

Figure 3.37 Complete implementation of head-tracked 3-D loudspeaker audio system, consisting of binaural synthesizer, high-frequency shelving filters, crosstalk canceller, head tracker, and reverberator.

HRTFs depends on the particular implementation of crosstalk canceller. After binaural synthesis, the individual binaural signals are processed with the high-frequency shelving filters (G_{iL}, G_{iR}). For simplicity, we have included separate shelving filters for each channel; in practice, we would use the efficient implementation described in the previous section. The high-frequency adjusted binaural signals are summed to a single binaural pair which is input to the crosstalk canceller. The crosstalk canceller may be implemented using either the recursive topology of figure 3.27 or the feedforward topology of figure 3.29. The output of the crosstalk canceller is sent to the loudspeakers. As described earlier, the parameters of the crosstalk canceller, binaural synthesizer, and shelving filters depend on the current head position. This dependency is indicated in figure 3.37 by connections with dashed lines.

Figure 3.37 also includes reverberation processing suitable to achieve control of perceived source distance. Prior to binaural synthesis, each source is scaled by a gain g_i intended to simulate the attenuation of direct sound due to air propagation. The unscaled sources are summed and fed to a reverberator that outputs a binaural signal. The circuit allows the direct-to-reverberant ratio of each source to be controlled by the scaling gains g_i, which provide independent distance control for each source. This method is fairly primitive; better methods for integrating artificial reverberation into a spatial auditory display have been described (Jot, 1996, 1997; Gardner, 1998).

4 PHYSICAL VALIDATION

In this chapter we present data verifying that the bandlimited crosstalk cancellers presented in the previous chapter are in fact effective at cancelling crosstalk at frequencies below 6 kHz and that the equalization zone of the crosstalk canceller can be steered using the described methods. The data also serve to quantify the performance of these systems in terms of objective physical specifications. Two evaluation methods are used: acoustical simulation and acoustical measurement.

The effectiveness of a crosstalk canceller can be determined by simulating the acoustics of the listening situation. This is quite easy to do if we know the acoustical transfer matrix (defined in equation 3.14 on page 48), which describes how the two outputs of the crosstalk canceller are transformed into acoustical pressures at the left and right ears of a listener. The acoustical transfer matrix depends on the individual details of the listener, the speakers, and the listening environment. It is useful to start with an idealized listening situation to determine the best possible performance for a given crosstalk canceller implementation. We will present results of simulating a variety of crosstalk canceller implementations under ideal conditions, where both the listener head model and the crosstalk canceller head model are based on KEMAR HRTFs, thus simulating the situation of an individualized crosstalk canceller. We will also use the simulations to show the spatial extent of the equalization zone, to demonstrate the validity of the steering methods, and to show how crosstalk cancellation is affected when the listener is displaced from the target equalization zone.

In addition to the acoustical simulations, a number of acoustical measurements of a crosstalk cancellation system were performed. The measurements were made using both the KEMAR dummy head microphone and also miniature microphones inserted into the ear canals of human subjects. KEMAR measurements were made in anechoic conditions in order to validate the acoustical simulations. Both KEMAR and human measurements were made in reverberant listening conditions, the same listening situation used for the psychoacoustic experiments described in the following chapter, and the humans used for recording are the same subjects in those experiments. The reverberant measurements of KEMAR can be compared to the anechoic measurements to quantify the effects of reverberation. The human measurements in

80   Physical Validation

reverberant conditions are particularly useful to quantify the effectiveness of crosstalk cancellers when used in real life situations.

The performance of a crosstalk canceller can be objectively described in terms of the frequency dependent *channel separation* at the ears of the listener. Channel separation must be calculated independently for left-to-right and right-to-left separation, although we expect that symmetrical listening situations will yield nearly identical channel separations for both sides. Channel separation is calculated by measuring the impulse response from each input of the crosstalk canceller to each ear, and then computing the magnitude of the left and right interaural transfer functions (ITFs), defined in equation 3.20 on page 50. When presenting audio via loudspeakers without a crosstalk canceller, assuming anechoic conditions, the naturally occuring channel separation is equal to the ITF magnitude corresponding to the incidence angle of the loudspeaker. We expect the channel separation to increase when the crosstalk cancellation is enabled, but this is not guaranteed.

4.1 ACOUSTICAL SIMULATION

Simulation of sound propagation from the speakers to the ears of a listener is easily accomplished by modeling the acoustical transfer matrix, defined in equation 3.14 on page 48. Our simulations assume ideal acoustical conditions, including flat speaker responses and an anechoic space. In addition, the listener's head transfer matrix is modeled using KEMAR HRTFs, and therefore the listening situation is individualized. Despite the individualized simulation, crosstalk cancellation will not be perfect when using a crosstalk canceller based on low-order filter approximations, i.e., filters intended for real-time implementation. However, the individualized simulations will yield better results than we expect for typical listeners.

This section will show some results of the acoustical simulations, including channel separation plots that compare different implementations of crosstalk cancellers, contour plots of acoustic equalization zones, and channel separation plots as the listener is moved away from the ideal listening location.

4.1.1 Channel separation for symmetric crosstalk cancellers

Figure 4.1 compares channel separation for various implementations of the symmetric crosstalk canceller. The plots show the frequency-dependent channel separation at the ears (solid line), obtained by computing the ITF magnitude resulting from crosstalk canceller inputs consisting of an impulse signal on the ipsilateral channel and a zero signal on the contralateral channel. Shown for comparison (dashed line) is the ITF magnitude for 30 degree incidence, i.e., the naturally occuring channel separation for sounds radiated from the loudspeaker. The difference between the dashed and solid lines represents the increase in channel separation due to the crosstalk canceller.

3-D Audio Using Loudspeakers

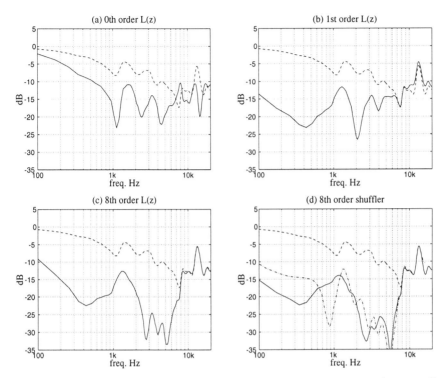

Figure 4.1 Comparison of channel separation for various symmetric crosstalk cancellers, implemented at 44.1 kHz sampling rate. In each plot, the dashed line is the natural head shadowing for 30 degree incidence and the solid line is the channel separation resulting after crosstalk cancellation. Plots (a) through (c) show results from a crosstalk canceller based on a parameterized head model, and plot (d) compares results from a non-parameterized shuffler filter implementation. (a) Head shadowing model implemented using delay and attenuation. (b) Head model implemented using delay, attenuation, and one-pole lowpass filter. (c) Head model implemented using delay and 8th-order IIR filter. (d) Shuffler filters implemented using 8th-order IIR filters; the dash-dot line is the 32 kHz implementation, used in many of the sound localization experiments.

Figure 4.1a shows the results for a crosstalk canceller based on the simplest possible head shadowing model, consisting of a frequency-independent delay and attenuation (Schroeder and Atal, 1966), used in the structure of figure 3.31, or equivalently figure 3.30. This head shadowing model corresponds to a 0th-order $L(z)$ filter, consisting only of a multiplicative term. The ITD was set to 0.25 msec (11 samples at 44.1 kHz) and the attenuation was set to 8.8 dB. These parameters were determined from the broadband ILD and ITD data for 30 degree incidence shown in figure 3.2 and figure 3.3, respectively. Although extremely simple, this crosstalk canceller is effective at increasing channel separation up to 7 kHz.

The broadband ITD and ILD values, which are essentially averaged across frequency, are substantially dependent on the value of the ITF at frequencies around 2-3 kHz, where the HRTF response has the most power due to the ear canal resonance. Consequently, basing the head shadowing parameters on broadband interaural differences yields a crosstalk canceller that is primarily effective at frequencies above 1500 Hz, in particular because the ITD corresponds to the high-frequency head diffraction model, discussed in section 3.2.5. We can greatly improve performance below 1500 Hz by increasing the ITD parameter by a factor of 3/2 to 0.38 msec and decreasing the attenuation to 1.5 dB, as suggested by the frequency dependent interaural data shown in figure 3.8 and figure 3.9. However, this causes channel separation above 1500 Hz to greatly decrease. Clearly, a frequency-dependent approach is needed to model the transition in head diffraction that occurs at about 1500 Hz.

Figure 4.1b shows results for a crosstalk canceller based on Griesinger's head model described in equation 3.56 on page 70, consisting of a delay, an attenuation, and a first-order lowpass filter. The delay (ITD/$T - m$) was set to 0.2 msec (9 samples at 44.1 kHz), the attenuation was set to 1.5 dB, and the cutoff of the lowpass was set to 1 kHz. These parameters were determined through a calibration procedure described by Gardner (1995): the parameters were adjusted in order to maximally lateralize a white noise sound. This simple head model performs remarkably well, particularly at low frequencies. We note that the low-frequency performance of this model is better than the frequency independent model, in part because the one-pole lowpass filter has a frequency dependent phase response with increased phase delay at low frequencies, like a real head.

Figure 4.1c shows results for a crosstalk canceller based on the 8th-order head shadowing filter shown in figure 3.11 on page 43. This filter was obtained by applying Prony's method to a lowpass filtered ITF; the modeling delay m was 4 samples (0.09 msec at 44.1 kHz). This filter is not quite as good as the first-order filter for frequencies below 400 Hz, but is generally better from 2–6 kHz. Above 6 kHz the crosstalk canceller has no effect, i.e., there is no change in channel separation relative to natural head-shadowing, because the head shadowing filter $L(z)$ is lowpass filtered at a 6 kHz cutoff. We might expect that the difference between the solid and dashed lines in figure 4.1c should equal the ITF error shown in figure 3.11. However, the ITF error only considers a single crosstalk term, whereas the channel separation data additionally considers higher-order crosstalks. Consequently these data differ somewhat at low frequencies, where high-order crosstalks are more significant.

Figure 4.1d shows channel separation data for symmetric crosstalk cancellers based on 8th-order shuffler filters designed using the procedure described in section 3.4.5, equation 3.55 on page 69. The solid line shows the results for a 44.1 kHz implementation, and the dashed line shows the results for a 32 kHz implementation that was used for several of the sound localization experiments described in the next chapter. We might expect the 32 kHz filter to perform better than the 44.1 kHz filter, because the 8th-order filter has more leverage at lower sampling rates; however, the 44 kHz filter performs better, which we cannot explain. The 44.1 kHz results also differ from the results in figure 4.1c, particularly at frequencies below 300 Hz, but the overall performance of these crosstalk cancellers seems about the same.

All the channel separation data in figure 4.1 (as well as the ITF error data in figure 3.11) exhibit a peak at about 1500 Hz. This frequency corresponds to a wavelength that is comparable to the size of the head; therefore, a transition between long and short wavelength head-diffraction models occurs at about this frequency (Kuhn, 1977, 1987). It seems that the low-order filter models are unable to capture this transition behavior adequately, as shown in figure 3.12.

4.1.2 Contour plots of channel separation

The equalization zone can be visualized as a set of equal-separation contours in the region of space near the target equalization location. The contour plot is created by first evaluating average channel separation (the average of the left-to-right and right-to-left channel separation) at a set of head positions distributed on a two-dimensional grid around the target location; the contour plot is then generated from the resulting data. Figure 4.2 shows the geometry of the simulated listening situation used for creating the contour plots. The standard listening situation is simulated and channel separation is evaluated at 5 cm increments on a 50 cm square grid centered on the ideal listening location at the origin of the coordinate system.

The head transfer matrix is simulated using KEMAR HRTFs, which are sampled at 5 degree increments on the horizontal plane. Simulated HRTFs at intermediate angles are obtained by linearly interpolating between adjacent HRTFs. In general this is a poor method for interpolating HRTFs (see Jot et al., 1995, for alternative methods), but in this case it is acceptable because of the dense spatial sampling and because high-frequency (> 10 kHz) accuracy is not an immediate concern. The speakers are assumed to be ideal tranducers (omnidirectional with flat frequency response), and the air propagation is modeled as a 1/r attenuation and a variable delay implemented using a third-order Lagrangian interpolator (Laakso et al., 1996).

Figure 4.3a shows a contour plot generated using a symmetric crosstalk canceller based on the 44.1 kHz, 8th-order head shadowing filter, shown in figure 3.11 on page 43. The contour plot is based on channel separation integrated from 100 Hz to 6 kHz. The spatial extent of the equalization zone is rather small: 10 dB or greater channel separation is achieved in a region only about 10 cm wide and 30 cm long. The spatial extent of the equalization zone is greater along the front-back axis than along the lateral axis because lateral head translations create unequal path lengths to the speakers, which seriously degrades crosstalk cancellation.

We expect the spatial extent of the equalization zone to depend on frequency; lower frequencies, and hence longer wavelengths, should create larger equalization zones. Figure 4.3b shows the equalization zone evaluated from 100 Hz to 1000 Hz, and as

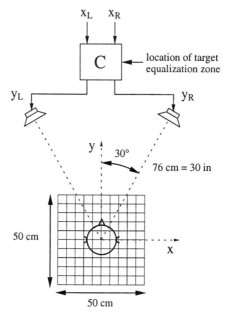

Figure 4.2 Geometry of simulated listening situation used to create channel separation contour plots. Channel separation is evaluated at each point on a 50 cm square grid of head positions, in 5 cm increments.

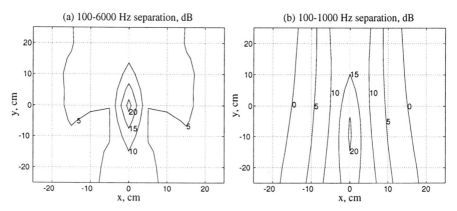

Figure 4.3 Contour plots of channel separation as a function of spatial location of the listener. The crosstalk canceller is based on an 8th-order head shadowing filter (at 44.1 kHz) and the equalization zone is steered to the ideal listening location at (0,0). Channel separation evaluated from (a) 100 Hz to 6 kHz and (b) 100 Hz to 1000 Hz. The spatial extent of the equalization zone is greater along the front-back axis than the lateral axis; this is especially true at lower frequencies.

3-D Audio Using Loudspeakers

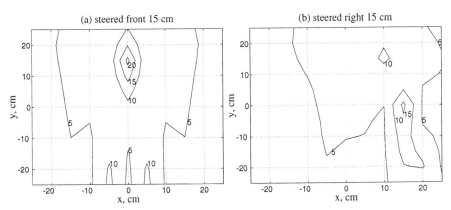

Figure 4.4 Plots of steered equalization zones using crosstalk canceller based on 8th-order head shadowing filter (at 44.1 kHz). Channel separation is evaluated from 100 Hz to 6 kHz. (a) Equalization zone steered 15 cm to the front by increasing the ITD parameter of the head shadowing model to 0.30 msec (nominally 0.25 msec). (b) Equalization zone steered 15 cm to the right by delaying the right speaker signal 0.43 msec and attenuating it by 1.6 dB.

expected, the spatial extent has increased considerably. The region of 10 dB or greater channel separation extends beyond the range of the plot in the front-back direction. It is well known that crosstalk cancellation systems are rather insensitive to front-back listener translations. This fact, along with the frequency dependence of the equalization zone, suggests that low frequencies are of paramount importance for crosstalk cancellation systems, a repeated theme in this document. Nevertheless, sound localization results in the next chapter will show that front-back translations do slightly degrade localization performance.

An interesting feature of figure 4.3b is that the equalization zone is displaced slightly to the rear of the origin. This suggests that the low-frequency channel separation in figure 4.1c would increase if the ITD of the head shadowing model, or equivalently the phase delay of the $L(z)$ filter, was increased. The data suggest that the 8th-order approximation to the ITF has insufficient phase delay at low frequencies, which is confirmed in figure 3.12 on page 44.

We now show plots of equalization zones that have been steered away from the ideal listening location. Figure 4.4a shows an equalization zone that has been translated 15 cm to the front towards the speakers. The steering was accomplished by increasing the ITD parameter of the head shadowing model to 0.30 msec (nominally 0.25 msec at the ideal listening location). The 8th-order head shadowing filter $L(z)$ was unchanged. It is clear that changing the ITD parameter alone can effectively steer the equalization zone along the front-back axis, at least within a local region.

Figure 4.4b shows an equalization zone that has been translated 15 cm to the right. The steering was accomplished by delaying and attenuating the right speaker signal;

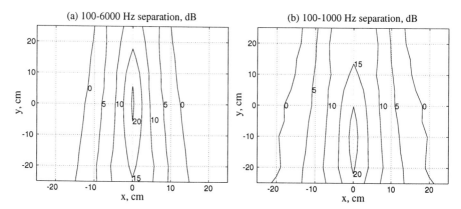

Figure 4.5 Contour plots of channel separation as a function of spatial location of the listener for loudspeakers positioned at ±5 degrees azimuth. The crosstalk canceller is a shuffler topology using 8th-order filters. Channel separation evaluated from (a) 100 Hz to 6 kHz and (b) 100 Hz to 1000 Hz. The results show that loudspeakers spaced at ±5 degrees yield a larger equalization zone than when spaced at ±30 degrees; however, the improvement is mostly along the front-back axis.

the delay was 0.43 msec and the attenuation was 1.6 dB. The attenuation has negligible steering effect compared to the delay, but was included for completeness. The plot shows that steering the equalization zone laterally reduces the maximum channel separation. The reason for this is that the listening situation is asymmetric; the left and right speakers are at different absolute incident angles because the listener is facing straight ahead. If the listener was rotated slightly left so that the absolute incident angles of the speakers were equal, the channel separation would be increased. In other words, delaying one of the outputs of a symmetrical crosstalk canceller to steer the equalization zone left or right is most effective when the listener faces the midpoint between the speakers.

The simulation technique allows easy manipulation of the listening geometry parameters. Of particular interest is the effect of different speaker spacings on the size of the equalization zone. Bauck and Cooper (1996) have proposed using closely spaced speakers as a way to widen the equalization zone; the rationale is that lateral head translations cause time-of-arrival differences between the two loudspeakers which are minimized by closely spaced loudspeakers. Figure 4.5 shows the simulated equalization zone for loudspeakers positioned at ±5 degrees azimuth. The crosstalk canceller was designed for 5 degree speaker azimuths and implemented using 8th-order shuffler filters (at 44.1 kHz). Figure 4.5a shows the equalization zone evaluated from 100 Hz to 6 kHz. Comparing these results to figure 4.3a shows that the 10 dB contour has widened by a few centimeters at y = 0 cm; however, the contour has grown considerably along the front-back axis. The channel separation from 100 Hz to 1000 Hz, shown in figure 4.5b, is not substantially different than the corresponding results for

3-D Audio Using Loudspeakers

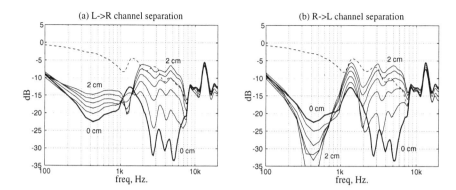

Figure 4.6 Plots of channel separation as head is laterally translated from target equalization zone at the ideal listening location: (a) left to right separation; (b) right to left separation. The data were generated using a crosstalk canceller based on an 8th-order head shadowing filter at 44.1 kHz. The bold line is the channel separation when the head is ideally located (labelled 0 cm), the solid lines show the channel separation as the head is moved 2 cm to the right in 5 increments, and the dashed line is the ITF magnitude for 30 degree incidence.

±30 degree loudspeakers, shown in figure 4.3b. The equalization zone widens only slightly when the speakers are positioned closely together.

4.1.3 Effect of lateral translation

Figure 4.6 shows how channel separation degrades when the head is laterally translated from the target equalization zone at the ideal listening situation: figure 4.6a shows the left-to-right channel separation; figure 4.6b shows the right-to-left channel separation. The data were generated using a crosstalk canceller based on the 8th-order head shadowing filter (at 44.1 kHz) discussed earlier. In the plots, the bold line is the channel separation when the head is ideally located (labelled 0 cm), the solid lines show the channel separation as the head is moved 2 cm to the right in 5 increments of 4 mm each, and the dashed line is the ITF magnitude for 30 degree incidence.

The plots show that crosstalk cancellation degrades considerably above 1 kHz for even small head displacements. This suggests that crosstalk cancellation is in general limited by the phase match between the acoustic crosstalk signal and the cancellation signal. It is interesting that the cancellation improves at certain freqencies as the head is translated; for instance, in figure 4.6a, cancellation improves at about 1300 Hz, whereas in figure 4.6b, cancellation improves at about 400 Hz. This phenomena indicates that the crosstalk cancellation filter accurately models these asymmetrical head responses in certain frequency ranges.

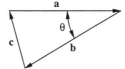

Figure 4.7 Phasor diagram of crosstalk cancellation.

Table 4.1 Relationship between cancellation k and phase error θ assuming crosstalk and cancellation signals are the same magnitude. Δd is positional error, corresponding to the phase error, for a 1 kHz frequency.

k, dB	θ, degrees	Δd, cm
0	60	5.7
-3	41	3.9
-6	29	2.7
-9	20	1.9
-12	14	1.4
-15	10	1.0
-18	7	0.7
-21	5	0.5

4.1.4 Phasor analysis

A simple phasor analysis shows why cancellation is so dependent on the phase match between the crosstalk and the cancellation signal. Let us represent the acoustic crosstalk signal and the cancellation signal as phasors **a** and **b**, respectively. Perfect cancellation results when **b** = -**a**, but usually there are phase and/or magnitude errors in **b**, so that cancellation results in a residual phasor **c** = **a** + **b**. Let us assume that the magnitudes of the two phasors **a** and **b** are equal, i.e., that the magnitude of the cancellation signal has been correctly chosen; this situation is depicted in figure 4.7. The cancellation is then related to the phase error θ by

$$k = \frac{|\mathbf{c}|}{|\mathbf{a}|} = 2\sin\left(\frac{\theta}{2}\right) \qquad (4.1)$$

where the cancellation k is defined as the ratio of the residual magnitude to the crosstalk magnitude. The residual is smaller than the crosstalk when $k < 1$; this can only occur when the phase error is less than 60 degrees, which corresponds to a positional error of 5.7 cm for a frequency of 1 kHz. In order to achieve 12 dB of cancellation, i.e. $k = 0.25$, the positional error must be less than 1.4 cm at 1 kHz. Table 4.1 summarizes the relationship in equation 4.1; positional errors are given for a 1 kHz frequency.

4.2 ACOUSTICAL MEASUREMENTS

We now describe acoustical measurements of crosstalk cancellers, first discussing KEMAR measurements made in anechoic conditions, and then discussing KEMAR and human measurements made in reverberant conditions.

4.2.1 Measurements of KEMAR in anechoic conditions

Channel separation resulting from crosstalk cancellation was measured using a KEMAR in MIT's anechoic chamber. The KEMAR was configured with model DB-066 and model DB-065 left and right "large" pinna, respectively. These two pinnae have similar sizes, but they are not perfectly symmetrical. The KEMAR was placed between two Cambridge SoundWorks[†] Ensemble satellite speakers, positioned at ±30 degrees at a distance of 76 cm (30 in). The crosstalk canceller measured was the 8th-order shuffler implementation at 32 kHz, which was used for several sound localization experiments, and whose ideal channel separation is shown in figure 4.1d.

Only the left-to-right channel separation was measured. As with the acoustical simulations, this was accomplished by measuring the impulse response of the left input channel of the crosstalk canceller to both ears. The left-to-right channel separation is then given by the magnitude of $ITF_L = H_{LR}/H_{LL}$. In addition to these measurements, the responses from each speaker to each ear (without the crosstalk canceller) were also measured.

Figure 4.8 shows the left-to-right channel channel separation. The results are disappointing; significant cancellation is only occuring below about 1.5 kHz. The data are similar to the head translation data of figure 4.6a. In fact, an inspection of the time response data reveals that the right speaker is several cm closer to the head than the left loudspeaker. The speaker to ear path lengths were carefully adjusted prior to the acoustical measurements; however, the floor in the anechoic chamber is a flexible wire mesh, and the KEMAR and speakers must have shifted after the experimenter left the chamber.

The individual speaker to ear measurements allow us to simulate the crosstalk cancellation acoustics using convolution, in a manner completely analogous to the acoustical simulations described earlier. The KEMAR crosstalk cancellation measurement shown in figure 4.8 was simulated using the speaker to ear responses and the results agreed closely with the true measurements, reassuring us that the loudspeakers, air propagation, head diffraction, and microphones are accurately modeled by linear, time-invariant filters.

[†]Cambridge SoundWorks, 311 Needham St., Newton, MA 02164.

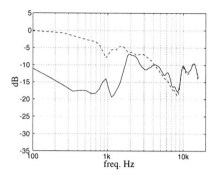

Figure 4.8 Left-to-right channel separation measured using a KEMAR in anechoic conditions. The solid line is the channel separation, the dashed line is the ITF magnitude for 30 degree incidence. The crosstalk canceller was based on 8th-order shuffler filters, at 32 kHz. The poor cancellation results above 1 kHz are due to a head position error.

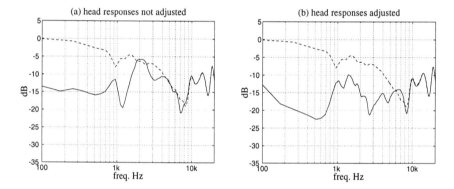

Figure 4.9 Left-to-right channel separation in simulation of crosstalk cancellation acoustics based on asymmetrical KEMAR measurements made in anechoic chamber (solid line): (a) head responses not adjusted for asymmetrical path lengths from speakers to ears; (b) head responses adjusted (delayed) to simulate ideal KEMAR positioning. High-frequency separation improves when KEMAR is correctly positioned. The crosstalk canceller was based on 8th-order head shadowing filter, implemented at 44.1 kHz. The dashed line shows natural head shadowing for 30 degree incidence.

In order to simulate a symmetrically positioned KEMAR, the head responses were appropriately delayed. The delays were determined by a cross-correlation analysis; the right speaker to right ear response was delayed by 0.052 msec (1.8 cm), and the right speaker to left ear response was delayed by 0.021 msec (0.7 cm). Figure 4.9 shows the results of simulating the acoustics using the crosstalk canceller based on an 8th-order head shadowing filter at 44.1 kHz, whose ideal channel separation was shown in figure 4.1c. Figure 4.9a shows the channel separation where the head responses have not been adjusted to create a symmetric listening situation, and figure 4.9b shows the adjusted situation. The non-adjusted results are of course similar to results in figure 4.8, whereas the adjusted results show an improvement in high-frequency separation as expected. The cancellation is relatively poor from 1-2 kHz; in this frequency range head shadowing is poorly modeled by low-order filters as shown in figure 3.12 and figure 4.1. In the next section we will also see evidence of asymmetric speaker responses at these mid-frequencies. Differences between the results of figure 4.9b and the ideal results of figure 4.1c are attributed to 1) asymmetries in the acoustical transfer matrix that are not corrected by delaying the head responses, such as the asymmetrical pinnae and possibly different speaker responses, and 2) differences between the crosstalk cancellation model, based on the "small" pinna, and the measurement KEMAR, which used the "large" pinnae.

Several conclusions can be drawn from these results. First, simulating head acoustics by convolution with measured head responses is accurate. Moreover, the data in figure 4.8 and figure 4.9 agree with the results of previous simulations, particularly in regards to how the cancellation degrades when the head is improperly positioned. These findings tend to validate our acoustical simulations. Finally, the simulation results of figure 4.9, based on asymmetrical measurements of KEMAR, show that cancellation is not greatly affected by slight asymmetries between the pinnae and the loudspeakers, except perhaps at mid-frequencies, but is greatly affected by asymmetrical path lengths from the speakers to the ears.

4.2.2 Measurements of humans and KEMAR in reverberant conditions

HRTF measurements of human subjects and the KEMAR were made in a reverberant environment. The measurements were used to verify the performance of a crosstalk cancellation system. The human subjects, participants in the sound localization experiments described in the following chapter, were measured during the localization experiment sessions. Consequently, the listening situation, i.e., the room, audio equipment, listening geometry, etc., was the same as used for the localization experiments. The transfer function from each loudspeaker (at ±30 degrees) to each ear of each subject was measured. The KEMAR was similarly measured by placing it in the ideal listening location. The measured HRTFs allow the channel separation of the crosstalk cancellers to be evaluated, using the simulation procedures described earlier.

The human measurements were made at the entrance of the blocked ear canal. The microphones used were Sennheiser[†] KE 4-211-2; these were placed into modified "swimmer-style" polymer earplugs. The microphone assembly was inserted into each ear canal so that the microphone face was flush with the ear canal entrance. The earplug acoustically blocked the ear canal and provided a stable, yet comfortable fit. The two microphones used had extremely well matched responses.

[†]Sennheiser Electronic Corp, P.O. Box 987, Old Lyme, CT 06371.

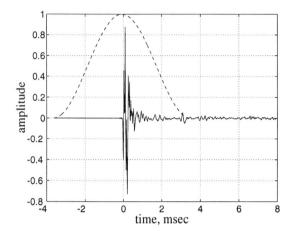

Figure 4.10 Time response of ipsilateral ear measurement (solid line) and Hanning window chosen for analysis (dashed line). The time axis is shifted so that time 0 corresponds to the initial sound arrival.

Measurements were made using the MLS technique described earlier. The sequence length was 16383 samples (370 msec at 44.1 kHz); this was sufficient to sample the room reverberation without significant time aliasing (the 60 dB reverberation decay time was 250 msec at 500 Hz).

Figure 4.10 shows a plot of a typical ipsilateral ear head response; there is an initial gap of 3.6 msec which includes air propagation plus a 1.1 msec inherent delay in the measuring system, followed by the head response and then the room reverberation. A geometrical analysis of the listening situation indicates that we should expect discrete early reflections from the opposite speaker at 2.5 msec, from the ceiling at 3.2 msec, from the rear wall at 4.8 msec, and from the floor at 6 msec. The ceiling reflection at 3.2 msec is quite prominent. We have chosen to window the data using a 7.2 msec Hanning window centered on the start of the head response as shown in figure 4.10. The window extends 3.6 msec forward and thus overlaps the ceiling reflection, but the reflection is highly attenutated by the window. In fact there is a continuum of echoes after the initial response, caused by reflections off the speaker stand and the bar connecting the speakers. The bar is used to mount apparatus described in the following chapter.

HRTF measurements were obtained for seven human subjects (B, C, D, E, F, G, and H, as described in the following chapter). Figure 4.11 shows a comparison of the measured ITFs and the KEMAR ITF, used for our crosstalk cancellation head model, whose ITF has been described previously in section 3.2.5. Figure 4.11a compares the ITF magnitudes; the thick solid line is the KEMAR ITF and the dotted lines are the subjects' ITFs. The subjects' ITF magnitudes are less than the KEMAR, especially for lower frequencies; however, this is due in part to the windowing. Using longer

3-D Audio Using Loudspeakers

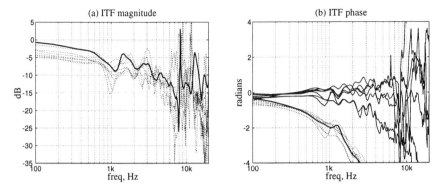

Figure 4.11 Comparison of measured ITFs of human subjects and KEMAR ITF, for right-to-left incidence. (a) ITF magnitude: subjects (dotted lines), KEMAR (thick solid line). (b) ITF phase: subjects (dotted lines), KEMAR (thick solid line), phase difference between subjects and KEMAR (solid lines). The two subjects with the largest heads have the largest phase differences.

windows increases the low-frequency convergence between the subject and KEMAR HRTFs, but it also admits more reflections which have a distorting influence. It should be emphasized that the ITF data are highly dependent on the choice of window shape and size.

Figure 4.11b compares the ITF phases: The unwrapped interaural phases of the subject measurements are shown with dotted lines; the KEMAR interaural phase is shown with a thick solid line. As described in section 3.2.5, above 1500 Hz the interaural phase delay is approximately constant; this corresponds to a linear phase term which appears as an exponential on a log frequency axis. The thin solid lines in figure 4.11b plot the differences between the KEMAR interaural phase and the subjects' interaural phases. Because of differences in head size, and thus ITD, the phase differences diverge exponentially at higher frequencies. For some subjects, the phase differences above 2 kHz are substantial; recall from phasor considerations that a phase difference less than $\pi/2 = 1.5$ radians is necessary for the cancellation process to decrease the magnitude of the cancelled signal.

The ITDs of the subjects were determined by linear regresssion on the interaural excess phase, and the results are given in table 4.2. The left-to-right and right-to-left ITDs were averaged to yield the ITDs given in the table. The ITDs were then used to calculate a spherical head model diameter D, also given in the table. The mean ITD is 0.237 msec (0.025 msec standard deviation), and the mean D is 15.9 cm (1.7 cm standard deviation). The data agree with the author's subjective impression of the subjects' head sizes. The data also correspond exactly to the phase difference plots in figure 4.11b; the two subjects with the largest head size, C and F, have the most negative ITF phase difference (the lowest plots), as expected.

Table 4.2 Interaural tine delay (ITD) and corresponding spherical head diameter (D) for human subjects measured at 30 degrees incidence. ITDs are the average of the left-to-right and right-to-left ITDs, calculated by linear regression on the interaural excess phase (c = 344 m/sec).

Subject	ITD, msec	D, cm
B	0.234	15.7
C	0.267	18.0
D	0.242	16.3
E	0.238	16.0
F	0.267	18.0
G	0.207	13.9
H	0.205	13.8

Figure 4.12 shows how crosstalk cancellation improves channel separation for the measured subjects, using the crosstalk canceller based on an 8th-order shuffler filter (at 32 kHz), whose ideal performance is given in figure 4.1d. This crosstalk canceller was used for some of the localization experiments described in the next chapter. The crosstalk cancellation acoustics were simulated using the time windowed HRTF data. Figure 4.12a shows the average channel separation using a crosstalk canceller (solid line), and the average ITF magnitude at 30 degrees (dashed line); the averages are superimposed on the individual subject data (dotted lines). Figure 4.12b shows the average increase in channel separation (solid line) superimposed on the individual subject data (dotted lines). The data show that crosstalk cancellation improves channel separation up to the cutoff frequency of 6 kHz. Note that there is significant variation between subjects at high frequencies, principally due to ITD differences, and for some subjects crosstalk cancellation reduces channel separation, even for frequencies as low as 1 kHz. Nevertheless, below 1 kHz there is significant improvement in channel separation.

Although not labeled in the plots, the subjects with the poorest high-frequency performance are the subjects with the largest head size, subjects C and F. As predicted by figure 4.11b, the poor cancellation is due to phase errors resulting from ITDs that differ greatly from the crosstalk cancellation head model. This suggests that the ITD parameter should be incorporated into a non-individualized crosstalk cancellation system. The listener's ITD could be determined via the calibration task used by Damaske (1971), i.e., the listener could adjust the ITD in order to maximally lateralize a 2-6 kHz bandpass filtered noise. Interestingly, subjects C and F are among the three subjects who had poor performance at 1 kHz; the third subject, subject E, has a nearly average ITD. We speculate that the unusual results at 1 kHz are related to facial

3-D Audio Using Loudspeakers

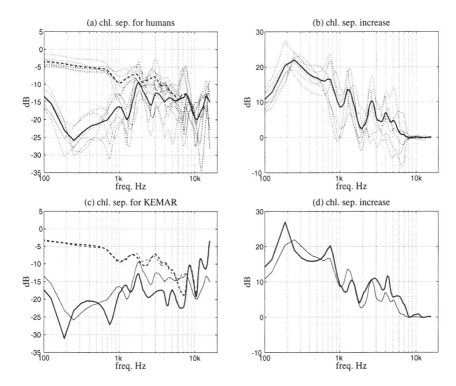

Figure 4.12 Right-to-left channel separation based on measurements of human subjects and KEMAR in reverberant conditions. (a) Channel separation for human subjects using crosstalk canceller (lower group of dotted lines), ITF magnitude at 30 degrees (upper group of dotted lines), and corresponding averages (thick solid and dashed lines, respectively). (b) Increase in channel separation for human subjects (dotted lines), and average increase (thick solid line). On average, crosstalk cancellation increases channel separation up to the cutoff of 6 kHz. (c) Channel separation for KEMAR using crosstalk canceller (thick solid line) and KEMAR ITF magnitude at 30 degrees (thick dashed line) superimposed on average human data from (a). (d) Increase in channel separation for KEMAR (thick solid line) superimposed on average human data from (c). The KEMAR results are not substantially better than the average human, probably due to the limiting influence of reflections and other asymmetries.

shape; it was noticed that these three subjects have relatively flat faces and wide cheekbones.

Figure 4.12c shows the channel separation for the KEMAR HRTF measurements made in the reverberant room, using the crosstalk canceller (thin solid line), and without the crosstalk canceller (thin dashed line), superimposed on the average human results of figure 4.12a (thick solid and dashed lines, respectively). Figure 4.12d shows the increase in channel separation for KEMAR (thin solid line), superimposed on the average human result from figure 4.12b (thick solid line). Because the

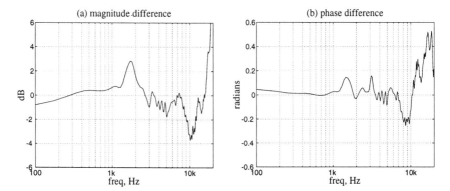

Figure 4.13 Response differences between the two loudspeakers: (a) magnitude difference; (b) phase difference. The 3 dB magnitude difference near 2 kHz accounts for the poor cancellation performance at that frequency.

crosstalk canceller is based on KEMAR, we expect the KEMAR to perform significantly better than the typical human. Indeed, KEMAR does perform better at high frequencies, but overall the performance is not substantially better than the average human performance. This is most likely due to both the limiting effects of room reflections captured within the time analysis window and also differences between the KEMAR's large pinna and the small pinna used for the crosstalk cancellation head model.

Particularly troubling is the relatively poor performance at frequencies near 2 kHz; this is seen in both the average human results and the KEMAR results. Further analysis revealed that the poor performance in this band is due to asymmetries in the ipsilateral head responses. The responses of both speakers were measured using an omnidirectional microphone placed at the ideal listening location. The differences between the magnitude and phase responses of the speakers are shown in figure 4.13. The time responses were windowed as shown in figure 4.10. Although the phase responses of the speakers are extremely well matched, the magnitude responses differ by nearly 3 dB near 2 kHz. It is not clear if this is due to the loudspeakers or to nearby reflections, though the latter seems unlikely because the speaker stands and other nearby apparatus are fairly symmetrical. We conclude that the mismatch in ipsilateral responses accounts for the poor cancellation at 2 kHz, and that this is caused by either a speaker response mismatch or an asymmetrical reflection effect. The mismatched responses may warrant the use of custom equalization filters per loudspeaker, though this was not done for our study.

Summarizing the previous sections, the following sources of crosstalk cancellation errors have been identified:
- Positional errors.

3-D Audio Using Loudspeakers

- Differences between the listener's ITF and the model ITF. This includes variation in ITD caused by head size differences.
- ITF modeling errors, i.e., errors caused by approximating a head model using low-order filters.
- Reverberation in the listening space.
- Asymmetries in the ipsilateral responses, including speaker response errors.

Much additional work would need to be done to exactly quantify these effects.

5 PSYCHOPHYSICAL VALIDATION

This chapter describes a set of sound localization experiments that test the effect on sound localization performance of tracking the listener's head position and using this information to optimize the acoustical presentation. We expect such optimization to increase localization performance, relative to the untracked condition, for the following reasons:

- Steering the equalization zone to the listener's position increases the low-frequency channel separation at the ears. Therefore, the intended binaural cues, particularly the low-frequency ITD cue, are more faithfully delivered.
- Head position is used to adjust the high-frequency powers to deliver an optimal high-frequency ILD cue.
- Source locations are maintained relative to the current head position so that a stationary external scene is synthesized. This enables dynamic localization cues during head motion.

We will use the term "tracked" to indicate the condition in which the audio system is compensating for the tracked position of the listener's head, and "untracked" to indicate that the audio system is not compensating for the listener's position[†]. The untracked condition is equivalent to the tracked condition when the listener is positioned at the ideal listening location.

Two initial experiments tested localization of synthetic sources using headphones and loudspeakers under ideal conditions. These experiments were followed by additional loudspeaker experiments that compared localization under tracked and untracked conditions. Experiments were conducted sequentially in separate sessions as features were added to the loudspeaker audio system. In order, the following experiments were conducted:

1. Headphone presentation, fixed head.

[†]These terms are equivalent to the "steered" and "unsteered" terms used in (Gardner, 1997).

2. Loudspeaker presentation, listener in ideal position, fixed head.

3. Loudspeaker presentation, listener translated laterally, fixed head, tracked and untracked conditions.

4. Loudspeaker presentation, listener translated to the front and to the rear, fixed head, tracked and untracked conditions.

5. Loudspeaker presentation, listener rotated, fixed head, tracked and untracked conditions.

6. Loudspeaker presentation, listener's head rotating right during stimulus presentation, tracked and untracked conditions.

With few exceptions, the subjects and experimental protocol varied little between experiments. One important difference is that only experiments 1 and 2 tested a full sphere of target locations and elicited both azimuth and elevation judgements. The remaining experiments only test horizontal target locations; elevation judgements were not obtained for these experiments. Experiments 1 and 2 were originally intended to yield baseline results for the remaining experiments. However, these experiments do not serve as true baseline conditions because of the increased range of the targets and allowable responses. This causes a slight increase in response variation relative to the experiments that only test horizontal locations.

All of the experiments use binaural stimuli synthesized using KEMAR HRTFs, but the details of the signal processing differ for each experiment. The experiments were conducted in parallel with the development of the steerable crosstalk cancellers discussed in Chapter 3. After each new steering capability was developed, it was tested. Consequently, each loudspeaker experiment uses a different crosstalk canceller, usually the simplest circuit that would accomplish the required steering task. In addition, the signal processing for the headphone experiment differs substantially because it does not require a crosstalk canceller. The experiments, signal processing, and results are discussed in detail in the following sections.

5.1 HEADPHONE AND LOUDSPEAKER LOCALIZATION EXPERIMENTS

This section describes the initial experiments that tested localization of synthetic sources presented over headphones and loudspeakers. We will describe the experimental procedures for each experiment and then discuss the results from both experiments together.

5.1.1 Headphone experiment procedure

The stimuli were created by processing a monophonic source sound with a binaural synthesizer and were presented to the listener over headphones. The source was a set of 5 pink noise bursts, 250 msec in duration with 10 msec linear onset and offset

3-D Audio Using Loudspeakers

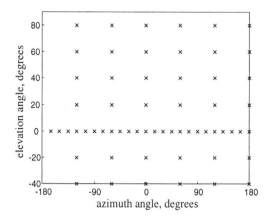

Figure 5.1 Set of all target locations used for experiments 1 and 2. On the horizontal plane, targets are spaced in 15 degree azimuth increments. At other elevatations, targets are spaced in 60 degree azimuth increments.

ramps, with 500 msec gaps between bursts. The source was processed to encode directional cues for a target location and was presented to the subject over headphones at a listening level of approximately 70 dBA SPL. Binaural synthesis was accomplished by filtering the source with KEMAR diffuse-field equalized HRTFs, sampled at 32 kHz. The convolution was accomplished using 128-point FIR filters. The resulting binaural signal was presented to the subject over AKG-K240 circumaural headphones. Moller et al. (1995a, 1995b) determined that the AKG-K240 DF headphones were approximately diffuse-field equalized for a typical human. We have used the same measurement technique to show that the AKG-K240 headphones used for the experiments are also approximately diffuse-field equalized.

Each sound localization trial was the same. A target location was randomly chosen from a set of 60 possible locations as follows: 24 locations on the horizontal plane in increments of 15 degrees azimuth, and 6 locations each at -40, -20, +20, +40, +60, and +80 degrees elevation in increments of 60 degrees azimuth. The set of target locations is shown in figure 5.1. Each location was tested exactly once per subject.

There was no attempt to control target distance. The synthesis HRTFs were measured at 1.4 m distance, so we might consider this to be the target distance, but there is no evidence that usable distance cues are incorporated in HRTFs measured this far from the head. Stimuli were presented at a constant level and artificial reverberation was not added, so loudness and reverberation cues to distance were not present.

The subject was instructed to report the perceived location of the auditory image by verbally reporting the azimuth and elevation angles and distance from the head. Two clock face diagrams showing azimuth and elevation angles and a distance diagram were placed in front of the subject to aid the reporting process. These diagrams are

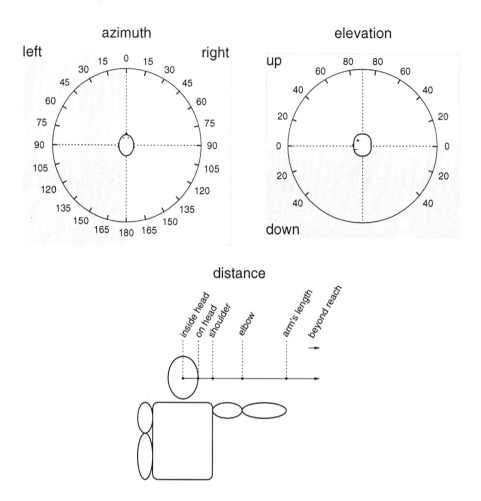

Figure 5.2 Azimuth, elevation, and distance charts used in the sound localization experiments.

shown in figure 5.2. Azimuth angles were reported in 15 degree increments, from 0 to 180 degrees, with "left" or "right" indicating the left or right hemisphere. Elevation angles were reported in 20 degree increments from "down 40" to "up 80." Distance judgements were given in a somatocentric coordinate system: positions inside the head (closer to the center than the surface) were reported as "in head," positions close to the surface of the cranium were "on head," and external locations were "shoulder" length, "elbow" length, "arm" length (i.e. reachable with the hand), or "beyond reach." Subjects reported azimuth, elevation, and distance judgements in any order. Many subjects found that it was easiest to make a fist at the perceived location of the image (if possible) and then report this position. Subjects were instructed to keep their head still during presentation of the stimulus.

3-D Audio Using Loudspeakers

Nine paid subjects (3 female, 6 male) volunteered for testing. We will refer to the subjects as A, B, C, D, E, F, G, H, and J. Subjects E, F, and J were female. Eight of these subjects (A, B, C, D, E, F, H, and J) participated in the headphone experiment. Each subject was tested once. None of the subjects had any prior experience with 3-D audio systems or sound localization experiments, and none reported any known hearing loss.

Prior to the session, subjects were given a brief demonstration and explanation of the 3-D technique. A helicopter sound was slowly panned completely around their head at 0 degrees elevation, and again at -40 and +40 degrees elevation. The sound was also panned from -40 to +90 degrees elevation at 90 degrees right azimuth. The panning was accompanied by an explanation of where the sound was supposed to be coming from. Most subjects reported that it was impossible to hear the sound in front of their head. This panning procedure was repeated with the pink noise bursts. The demonstration can be considered a form of training with feedback. However, subjects were encouraged to honestly report the perceived location of sound, and not to second-guess the experiment.

After this introduction, the response coordinate system was explained, and a set of ten training trials was conducted to familiarize the subject with the trial and response procedure. Feedback was not given. Following the training session, the set of 60 trials was conducted in two groups of 30 trials separated by a short break. The order in which trial locations were tested was random and differed for each subject. Subjects were completely unaware of the set of trial locations being tested. The experimenter (the author), present during the test, was aware of the set of locations being tested, but unaware of the presentation order.

5.1.2 Loudspeaker experiment procedure

The loudspeaker experiment procedure was similar to the headphone experiment, differing only in stimulus creation and the positioning of the subject's head. The stimuli were created by processing a monophonic source sound with a binaural synthesizer and a crosstalk canceller, and were presented to the listener over loudspeakers. The source sound was the same set of pink noise pulses used for the headphone experiment. The loudspeaker signals were created by combining a binaural synthesizer with a symmetric crosstalk canceller based on a shuffler topology, shown in figure 3.23 on page 57. Binaural synthesis was accomplished by filtering the source with KEMAR HRTFs, sampled at 32 kHz. The HRTFs were free-field equalized with respect to 30 degree incidence, and implemented using 128-point FIR filters. The resulting binaural signals were processed by the shuffler crosstalk canceller. The shuffler filters were designed from KEMAR HRTFs using the method described by equation 3.55 on page 69. The incidence angle was 30 degrees, and the crosstalk cancellation cutoff frequency was 6 kHz. The shuffler filters were implemented using 8th-order IIR filters. High-frequency compensation, as described in section 3.4.6, was not implemented; it has negligible effect when the listener is symmetrically positioned.

The experiments were conducted in a sound studio[†] (MIT room E15-485a) with dimensions of 4.7L x 4.3W x 2.1H m (15.5L x 14W x 7H feet), and a 500 Hz reverberation time of approximately 230 msec. With respect to the listener's orientation, the front and right-hand walls are more absorbent than the rear and left-hand walls.

Two Cambridge SoundWorks Ensemble satellite speakers were used as the sound sources. The speakers were positioned 76 cm (30 in) from the center of the subject's head, at ±30 degrees azimuth and 0 degrees elevation with respect to the subject. A visual sighting aide, consisting of a mask with two vertical slits positioned in front of a point light source, was used to ensure proper subject head position (Theile and Plenge, 1977). This apparatus was mounted on a rigid beam spanning the two speaker stands. Subjects were seated in front of the speakers and instructed to position their heads so that the light was visible though each slit. At the ideal listening position, the light beams from the sighting aide are spaced 6 cm apart (equal to the average interoccular spacing of an adult) and have a width of approximately 4 mm. Slight lateral head motions or head rotations cause one of the beams to become occluded. Front-back and up-down head motions were much less constrained, but these were deemed to be less important to control.

The same sighting apparatus was used in following experiments when the subjects' heads were not in the ideal listening location. This was possible because the apparatus could be positioned anywhere along the beam connecting the two loudspeakers, angled to point in different directions, and the distance between the point light source and the slits could be adjusted to accomodate different distances between the subject and the sighting apparatus.

The loudspeaker experiment differed from the headphone experiment in the following ways:

1. The listening level was set to 64 dBA SPL, measured at the center of the listener position using continuous pink noise played through one speaker. The levels of both speakers were set this way to balance the channel amplitudes. The SPL was set lower than the headphone experiment (approximately 70 dBA) in an effort to achieve the same subjective loudness.

2. Prior to the experiment, the height of the speakers was adjusted so that the center of each speaker face was aligned with the listener's ears. The subject was then trained in the sighting task used for head alignment. After the subject felt comfortable with the task, the subject was asked to stay sighted and the distance from each speaker to the tragion of the ipsilateral ear was measured. If the subject was not at the proper distance from the speakers, the chair would be moved and the

---

[†]The headphone experiment was conducted in the same room.

3-D Audio Using Loudspeakers

procedure repeated.

3. Prior to each experimental trial, the subject was asked whether he/she was ready. The subject then positioned his/her head and responded OK. The stimulus was played and the subject responded. As with the headphone experiment, subjects could consult the three charts shown in figure 5.2 to assist in reporting the location. These were placed on a music stand below the sighting apparatus.

4. Subjects A, B, C, D, E, F, G, and H participated in the loudspeaker experiment.

Prior to conducting trials, a scripted demonstration of the loudspeaker system was given, using the same procedure as the headphone demonstration. Then a set of 10 training trials was conducted without feedback, followed by the 60 localization trials.

5.1.3 Statistical analysis

Localization performance is subject to three principal types of angular errors:

- Systematic errors between the mean judged location and the target location that have the form of a response bias. Following Blauert (1983), we will call these errors "localization errors". In free-field listening, the smallest errors are seen in azimuth judgements for frontal horizontal targets, and the largest errors are seen in elevation judgements for rear medial targets. In our experiments, we expect additional errors over free-field conditions due to systematic variation between the synthesis and listener HRTFs, and also due to linear distortions in transmitting the binaural signals to the ears.

- Variation of the responses around the mean, attributed to perceptual noise. Blauert (1983) defines the "localization blur" to be the amount of displacement of the target that is recognized by 50% of the listeners as a change in judged location, in other words the just noticeable difference (JND). Psychophysical models relate the JND to the probability distribution of the responses (Durlach, 1968). Rather than use the strictly defined term localization blur, we will call these errors "response variation".

- Front-back and up-down reversals (also called confusions), where a target location is confused with the mirror symmetric location obtained by reflecting the target across the frontal plane (for a front-back reversal) or the horizontal plane (for an up-down reversal). Compared to front-back reversals, which are common, up-down reversals (Wenzel et al., 1993) are less common and difficult to distinguish from the other types of errors.

In addition to these errors, we also expect variation in the distance responses. As explained earlier, no attempt was made to control target distance. We will report the judged distances without consideration of a target distance.

Following Wightman and Kistler (1989b), we will use the following statistics to characterize the results: front-back reversal counts, average angular error, and inverse kappa (κ^{-1}). A judgement is considered to be a front-back reversal if the reversed judgement, obtained by reflecting the judgement across the frontal plane, is closer to the target. Front-back reversals are separately classified as front-to-back (F→B) reversals and back-to-front (B→F) reversals (Wenzel et al., 1993). Front-back reversals are counted and corrected before further analysis. Unlike Wenzel et al. (1993), we do not correct up-down reversals, because we believe most elevation judgements that would be classified as up-down reversals are actually the result of localization error or response variation.

The average angular error is the mean unsigned angle, i.e., as measured on a great circle, between each corrected judgement vector and the corresponding target vector. κ^{-1} is a statistic for spherically distributed data that characterizes the spread of judgements around the judgement centroid (Fisher et al., 1987; Wightman and Kistler, 1989b). Analogous to variance, κ^{-1} is small when the judgements are tightly clustered, and is large when judgements are highly dispersed.

Experiments 1 and 2 test a full sphere of target locations, but all the remaining experiments test only horizontal targets, and elevation judgements were not elicited. We can compare the results from experiments 1 and 2 to the subsequent experiments by considering the azimuth judgements in response to horizontal targets. We will analyze the azimuth judgements in terms of the lateral (LR) angle, i.e. the angle subtended by the judgement vector and the median plane (called a "right-left" angle by Kistler and Wightman, 1992). The lateral angle is unaffected by front-back reversals, which are counted separately as previously described. When computing a lateral angle error, the target must also be converted to its corresponding lateral angle. There is little difference between using lateral angles or front-back corrected azimuth angles; both yield the same angle error statistics. We prefer plots of the judged lateral angle when localization performance is severely degraded.

5.1.4 Results

Figure 5.3 shows histograms of judged azimuths at each target azimuth on the horizontal plane, across all subjects, for both headphone and loudspeaker presentation. Error-free localization would result in a straight line of responses along the $y = x$ diagonal. The histograms clearly show both response variation and front-back reversals. With headphones, almost all of the target locations are perceived in the rear. With loudspeakers, most of the front targets are correctly perceived in front, but many of the rear targets are also perceived in the front.

Front-back reversal percentages for horizontal targets and all targets are given in Table 5.1 on page 109. The pattern of front-back reversals is very specific to the individual subject; this is shown in figure 5.4, which is a bargraph of the individual reversal percentages for horizontal targets. With headphones, only subject D reversed a

3-D Audio Using Loudspeakers 107

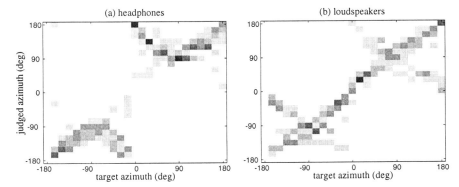

Figure 5.3 Histograms of judged azimuth at each target azimuth over headphones (a) and loudspeakers (b), all subjects. All targets are on the horizontal plane. White indicates no responses, black indicates 100% response frequency. The histograms clearly show that a great deal of front-to-back confusions occur for the headphone system, and the loudspeaker system images frontal locations better.

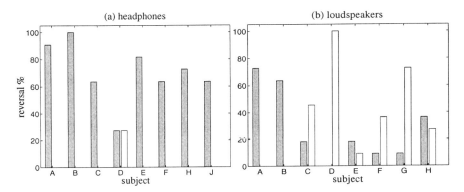

Figure 5.4 Bargraphs of individual front-to-back (gray bars) and back-to-front (white bars) reversal percentages, over headphones (a) and loudspeakers (b). All targets are on the horizontal plane. Over headphones, most frontal locations are reversed to the rear and few rear locations are reversed to the front. Over loudspeakers, the total reversal rates are about equal. In both cases the pattern of reversals is specific to the individual.

rear location to the front, and subject D had the lowest percentage of front-to-back reversals. With loudspeakers, Subect D reversed all rear targets to the front, and reversed none of the frontal targets. Subject D clearly has a propensity to perceive the stimuli as frontal. Subjects A and B, on the other hand, have a propensity to perceive the stimuli as from the rear.

Figure 5.5 plots the mean judged lateral angle (labeled LR angle) resulting from horizontal targets, averaged across all subjects, for both headphone and loudspeaker pre-

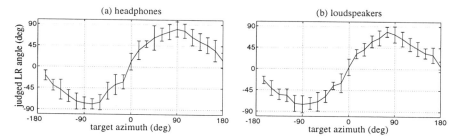

Figure 5.5 Mean judged lateral angle, averaged across all subjects, over headphones (a) and loudspeakers (b). All targets are on the horizontal plane. Errorbars show ±1 standard deviation. In terms of lateral angle, localization performance is similar over headphones and loudspeakers.

sentation. The plots are fairly similar. Any variation in the responses for targets near ±90 degrees leads to a mean judgement that is biased towards 0, because of the definition of the lateral angle, and this bias at ±90 degrees is visible in the plots. The bias of the mean does not affect the average angle error statistic. These plots are shown to allow comparison with later experimental results.

Table 5.1 summarizes the error statistics for the headphone and loudspeaker experiments. The average angle error, front-back reversal percentages, and inverse kappa are listed. The average angle errors are similar for headphones (34.2°) and loudspeakers (32.4°). Wightman and Kistler (1989b) report average angle errors of 19.1°, 18.8°, and 29.1° for low, middle, and high elevations, respectively, when the stimulus is synthesized using individualized HRTFs and delivered over headphones, which are similar to the errors reported for free-field listening in the same study. Wenzel et al. (1993) report average angle errors of about 23° and 29° for low and high elevations, respectively, when the stimulus is synthesized using non-individualized HRTFs, measured from the ears of a human who localizes well, and delivered over headphones. Thus, the average angle errors in our experiments are somewhat greater than those reported by Wenzel et al. (1993), which may indicate that the KEMAR HRTFs are not as effective for spatial synthesis as the HRTFs of a good human localizer. In our experiments, the average angle error is dominated by elevation errors, and consequently the errors are much smaller (14.3° for headphones, 12.1° for loudspeakers) when only azimuth judgements for horizontal targets are considered. We would expect even smaller errors if targets and judgements were restricted to the horizontal plane.

Wightman and Kistler (1989b) report very low front-back reversal rates using individualized HRTFs over headphones. They report average total rates of 7%, 6%, and 23% for low, middle, and high elevations, respectively, which are about twice the rates found with free-field listening in the same study. Wenzel et al. (1993) report an average of 50% front-to-back reversals and 12% back-to-front reversals using non-individualized HRTFs over headphones. These rates can be compared directly to our

3-D Audio Using Loudspeakers

Table 5.1 Average angle error, front-to-back (F→B) and back-to-front (B→F) reversal percentages, and inverse kappa for headphone and loudspeaker localization experiments. Results are shown for all locations and horizontal locations. †Horizontal angle error is calculated with the judged elevation set to 0 to facilitate comparison with later experiments.

	Avg. angle error	F→B	B→F	κ^{-1}
hdph: all locations	34.2°	78.5%	7.8%	0.14
hdph: horiz locations	14.3°†	70.5%	3.4%	n.a.
spkrs: all locations	32.4°	31.0%	46.6%	0.13
spkrs: horiz locations	12.1°†	28.4%	36.4%	n.a.

results of 78.5% front-to-back reversals and 7.8% back-to-front reversals over headphones. Although performance is worse in our study, both studies demonstrate that frontal images are difficult to perceive using non-individualized HRTFs over headphones. It would appear that frontal images are more difficult to perceive using the KEMAR HRTFs for spatial synthesis than using the HRTFs of a good localizer. This may also depend on the differing equalization methods used in the two studies; our study used diffuse-field equalized HRTFs delivered over approximately diffuse-field equalized headphones, whereas Wenzel et al. (1993) used HRTFs equalized using the headphone to ear canal transfer function of a single subject, the same subject used for HRTF measurements.

The inverse kappa statistics for our experiments (0.14 for headphones and 0.13 for loudspeakers) are quite large. Wightman and Kistler (1989b) report a mean κ^{-1} of 0.03, 0.05, and 0.10 for low, middle, and high elevations, respectively, using individualized HRTFs over headphones. These results are similar to the results for free-field listening in the same study. Wenzel et al. (1993) report a mean κ^{-1} of about 0.07 and 0.09 for low and high elevations, respectively, using non-individualized HRTFs over headphones. Our κ^{-1} is larger because it represents a between-subject variance, whereas the statistics reported in the other studies represent the mean of the within-subject variances. We cannot calculate a within-subject variance because we only gathered one response per target location per subject. These results are compatible with the hypothesis that between-subject variance is larger than within-subject variance.

The headphone and loudspeaker results differ in terms of the judged distances, with loudspeaker presentation yielding larger judged distances. This is easily visualized in figure 5.6, which shows polar histograms of judged distance versus *judged* azimuth in response to all horizontal target locations, for both headphones and loudspeakers.

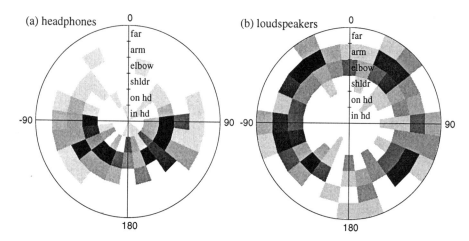

Figure 5.6 Two-dimensional polar histograms of judged distance as a function of judged azimuth for horizontal targets, over headphones (a) and loudspeakers (b), all subjects. White indicates no responses, black indicates maximum response frequency (5.2% for headphones, 4.2% for loudspeakers). Judged distances in the somatocentric coordinates are mapped to evenly spaced radii, labeled "in hd" (in head), "on hd" (on head), "shldr" (shoulder length), "elbow" (elbow length), "arm" (arm length), and "far" (beyond reach). The histograms don't show the mapping from targets to reponses, but just show the distribution of responses to a uniform distribution of horizontal targets. Frontal imaging and externalization is clearly better using loudspeakers.

These plots don't show the mapping from targets to judgements; rather, they show the distribution of judgements in response to a uniform distribution of horizontal targets. For headphones, the lack of frontal images is clear; also, images are not well externalized, tending to be concentrated at shoulder distance. The most distant images occur at lateral directions. In contrast, loudspeaker presentation shows excellent frontal imaging. Images are more externalized, tending to be clustered at arm's length. Rear images are less frequent, and tend to be perceived as closer to the head.

The dependence of judged distance on azimuth is shown in figure 5.7. The solid lines plot mean judged distance as a function of *target* azimuth, with ±1 standard deviation errorbars. The open diamonds plot mean judged distance as a function of *judged* azimuth, plotted only for azimuths where there are at least three responses. Thus, the diamonds show the means of the data plotted in figure 5.6.

The headphone data (figure 5.7a) clearly shows closer distance judgements than the loudspeaker data (figure 5.7b). Furthermore, the headphone data shows that judged distances depend on target azimuth, with medial locations localizing closer to the head. This phenomenon has also been reported by Begault (1992). For loudspeakers, the judged distances are relatively independent of target azimuth, but it is possible that the judged distances are dependent on *judged* azimuth. For instance, targets judged to be at ±150 degrees azimuth are judged closer to the head than the targets intended for

3-D Audio Using Loudspeakers

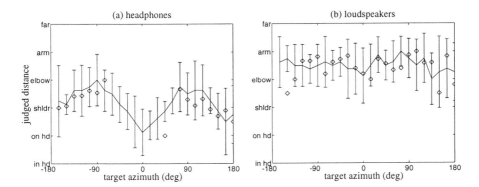

Figure 5.7 Mean judged distance as a function of target azimuth, over headphones (a) and loudspeakers (b), all subjects. Errorbars are ±1 standard deviation. Open diamonds plot mean judged distance as a function of judged azimuth, and are only shown for azimuths where there were at least three responses. Over headphones, images are localized closer to the head, particularly at medial target locations.

those locations. Further analysis shows that for these rear target locations, back-to-front reversals were accompanied by an increase in judged distance. Also, for frontal targets at ±30 degrees, front-to-back reversals were accompanied by a decrease in judged distance. We speculate that distance judgements are dependent on the resolution of targets to the front or rear hemispheres. Monophonic spectral cues influence both front-back resolutions (Blauert, 1969/70) and distance judgements (Hartmann and Wittenburg, 1996), which may explain the interdependence.

Figure 5.8 shows histograms of judged azimuth at each target elevation, with judged azimuth on the abscissa. These plots show the distribution of azimuth responses at each target elevation. Ideal performance would result in a response pattern identical to the distibution of target locations shown in figure 5.1 on page 101. Over headphones (figure 5.8a), there are few frontal responses, and it appears that targets at 40 degrees elevation were the most effective at generating frontal responses. Targets at 80 degrees elevation (almost overhead) are primarily localized at 180 degrees azimuth (behind). Over loudspeakers (figure 5.8b), azimuth judgements are more evenly distributed between front and rear locations. Targets at 80 degrees elevation (almost overhead) are primarily localized at 0 degrees azimuth (in front).

Figure 5.9 shows plots of mean judged elevation as a function of target elevation for non-medial target locations. The solid lines show the mean responses (with ±1 standard deviation errorbars) averaging across all subjects. The symbols show individual subject means at each elevation, where an open square is used for subject B, and crosses are used for all other subjects. At 0 degrees elevation, there are 22 non-medial locations, and at all other elevations there are only 4 non-medial locations (see figure 5.1 on page 101). Over headphones (figure 5.9a), performance is rather poor and there is a great deal of response variation. The mean judged elevation ranges

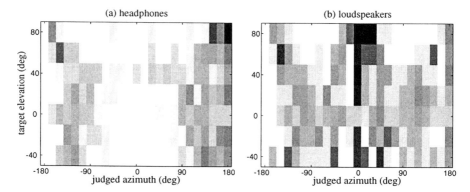

Figure 5.8 Histograms of judged azimuths at each target elevation, over headphones (a) and loudspeakers (b), all subjects. White indicates no reponses, black indicates maximum response frequency (38% for headphones, 21% for loudspeakers). Over headphones, high elevations tend to be localized in the rear, over loudspeakers, in the front. Over headphones, 40 degree target elevations yielded the largest percentage of frontal responses.

from -20 degrees to less than 40 degrees, and actually decreases for target elevations higher than 40 degrees. The breakpoint at 40 degrees elevation could be the result of the N1 feature disappearing from the HRTF spectrum at higher elevations (see figure 3.4 on page 33).

Combining the results in figure 5.9a and figure 5.8a, we see that high target elevations are perceived at low elevations close to the medial plane and in the rear. The large variation in the responses suggests that the just noticeable difference (JND) of elevation is very large, and therefore there are only a few response categories. Nevertheless, there are clearly some useful elevation cues in the stimuli.

Wenzel et al. (1993) report pronounced individual differences in elevation localization and this is also seen in our data. For instance, subject B's mean responses, indicated with open squares in the figure, are consistently above those of the other subjects. This could be the result of a reporting bias or the result of a systematic variation between the synthesis HRTFs and the listener's HRTFs. In Wenzel's study, some of the subjects localized elevations well and others poorly. In our study none of the subjects localized elevations particularly well. We speculate that this is because of differences between the synthesis HRTFs and the subject HRTFs, i.e. the KEMAR HRTFs do not closely match any of our subjects' HRTFs in terms of elevation features.

Figure 5.9b shows the same data for loudspeaker presentation. The mean judgements range from 0 to about 45 degrees, and the breakpoint at 40 degrees target elevation is not seen. Combined with the data in figure 5.8b, we see that high elevation targets are primarily localized at high elevations in the front. As with headphone presentation, subject B's responses are considerably higher than the mean. Overall, elevation per-

3-D Audio Using Loudspeakers

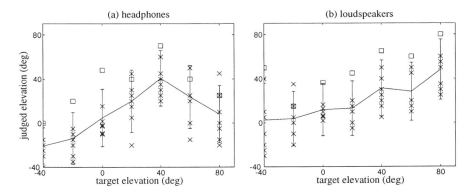

Figure 5.9 Mean judged elevation as a function of target elevation for non-medial target locations, over headphones (a) and loudspeakers (b), all subjects. Errorbars are ±1 standard deviation. Subjects' individual means at each elevation are shown with symbols: open squares for subject B, crosses for all other subjects. In both cases, elevation localization is rather poor and there is a great deal of response variation.

formance over loudspeakers appears to be poorer than over headphones, which is not surprising considering that the high-frequency cues are corrupted by crosstalk and room reverberation when delivered over loudspeakers.

Figure 5.10 shows the same elevation plots as figure 5.9, but for medial target locations. Because the synthesis HRTFs are perfectly symmetric, medial targets yield identical left and right binaural stimuli. The signals reaching the ear may not be identical because of asymetries in the transmission paths from the transducers to the ears; nevertheless, medial targets only encode a monophonic spectral cue for elevation localization. Over headphones (figure 5.10a), the mean judgements seem to be clustered near 0 degrees elevation, except for the target elevation of 40 degrees, which gave a mean judgement close to 40 degrees, and the 60 degree target, which gave a mean response of about 20 degrees. Because there are only two medial locations per elevation, we have omitted showing the individual subject means. Despite the huge variance in the responses, it would appear that there is a useful elevation cue in the 40 degree targets (which include both the front and rear medial locations) that is not present in the other targets. The data in figure 5.8a show that almost all of the infrequent frontal responses over headphones occured for 40 degree elevation targets, which suggests that the KEMAR HRTFs are more natural at this elevation in terms of monophonic spectral cues.

Over loudspeakers (figure 5.10b), the mean responses range from about 40 to 60 degrees, and the 40 degree target did not yield a considerable different mean response as with headphone presentation, though it did yield the highest mean response. The large elevation bias is striking, and may be explained by considering the constructive interference between the identical speaker signals. At the ears this results in reinforcement at frequencies where the period is a multiple of the ITD. For a speaker

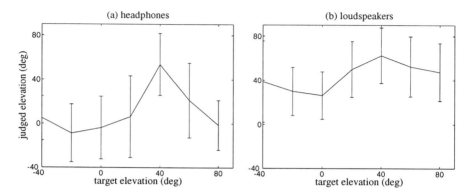

Figure 5.10 Mean judged elevation as a function of target elevation for medial target locations, over headphones (a) and loudspeakers (b), all subjects. Errorbars are ±1 standard deviation. Medial targets only encode monophonic spectral cues for elevation. Over headphones, it would appear the 40 degree target is encoding a useful elevation cue. Over loudspeakers, localization is particularly poor. The high elevation bias in the loudspeaker results can perhaps be explained by considering constructive interference of high freqencies at the ears.

angle of 30 degrees, the ITD is about 0.25 msec, and reinforcement occurs at multiples of 4 kHz. The fundamental at 4 kHz is compensated by the bandlimited crosstalk canceller, but higher harmonics above 6 kHz are not. The second harmonic at 8 kHz is close to the "overhead boosted band" at 9 kHz (Blauert, 1969/70; Hebrank and Wright, 1974b), and may bias the responses toward higher elevations[†]. The interference for medial target locations and a centered listener is not anticipated by the high-frequency strategy described in section 3.4.6, and is a deficiency of our approach.

We also notice that the non-medial results in figure 5.9 for target elevations of 80 degrees agree with the corresponding results in figure 5.10, which is not surprising considering that the 80 degree non-medial stimuli differ little from the 80 degree medial stimuli.

The results from the loudspeaker experiment can be compared qualitatively to two similar studies of 3-D audio loudspeaker systems by Damaske (1971) and Sakamoto et al. (1982). Damaske conducted experiments with a crosstalk canceller based on the interaural transfer function of a particular human subject. The ITF was obtained through a clever calibration task. Damaske's experiments showed excellent localization performance using speech stimuli recorded from a dummy head microphone and reproduced to fixed listeners in anechoic and reverberant rooms over two loudspeakers at ±36 degrees. Damaske did not detect and correct front-back reversals, and they appear in the data as increased response variation. Under anechoic listening condi-

---

[†]The second reinforcement harmonic occurs at 9 kHz exactly for an ITD of 0.22 msec, corresponding to a speaker angle of 25 degrees.

tions, using horizontal target locations, Damaske's results indicate far fewer front-back reversals than our results, and only back-to-front reversals occured. In slightly reverberant listening conditions (RT = 0.5 sec), the number of back-to-front reversals increased, and some front-to-back reversals occured. These results qualitatively agree with our results. Damaske separately tested elevation localization on the median plane, in anechoic listening conditions; his results are superior to ours. All medial directions including overhead and behind were perceived, and front-back reversals only occured for rear target locations that were reversed to the front. The anechoic listening conditions and the separate testing of medial locations clearly contribute to the increased medial performance seen in Damaske's experiments. It is also possible that the dummy head used supplied better medial localization cues than our KEMAR HRTFs.

In the experiments by Sakamoto et al. (1982), horizontal localization was tested in anechoic and reverberant conditions using speech stimuli presented over loudspeakers at ±30 degrees. The binaural synthesis and crosstalk cancellation were based on non-individualized HRTFs and combined into a "common" filter applied to both channels and a "ratio" filter applied to one channel. In anechoic conditions, localization performance with this system was nearly perfect. Front-back reversals only occurred for targets at 180 degrees, some of which were reversed to the front. In reverberant conditions (RT = 0.3 sec), the number of back-to-front reversals increased, and some front-to-back reversals also occurred, in a pattern very similar to Damaske's results and our results. Another experiment was conducted in anechoic conditions where the common filter was disabled, thus removing all monophonic spectral cues, and the results showed a dramatic increase in the number of back-to-front reversals.

The experiments by Koring and Schmitz (1993) are difficult to compare directly to our experiments because they primarily tested loudspeaker systems designed using individualized binaural synthesis and crosstalk cancellation. In anechoic and reverberant conditions, these systems performed far better than our non-individualized system. Experiments were also performed using a non-individualized system, but only in anechoic conditions, and results are only given for horizontal localization. The results generally agree with our results; back-to-front reversals were more common than front-to-back reversals, which only occured for medial (frontal) targets.

The non-individualized system of Koring and Schmitz (1993) was chosen by testing a number of different non-individualized systems (all based on human HRTFs) and selecting the one with the best overall performance, as determined by 10 test subjects. This procedure is essentially the same taken by Moller et al. (1996b) to select a "typical" set of HRTFs for binaural synthesis. We note that for the construction of a loudspeaker based audio system, the typical HRTFs must not only encode typical directional cues, but also have typical head diffraction properties for crosstalk cancellation.

5.2 VALIDATION OF HEAD TRACKING

This section describes sound localization experiments conducted using the loudspeaker system to validate the concept of steering the equalization zone to the location of the tracked head. As described in the chapter introduction, these experiments were:

3. Loudspeaker presentation, listener translated laterally, fixed head, tracked and untracked conditions.
4. Loudspeaker presentation, listener translated to the front and to the rear, fixed head, tracked and untracked conditions.
5. Loudspeaker presentation, listener rotated, fixed head, tracked and untracked conditions.

The experimental protocol for these experiments was similar to loudspeaker experiment 2. The principal differences are:

- Only horizontal target locations were tested. The subjects responded with azimuth and distance judgements only.

- The experiments randomly mixed trials between head-tracked and non-head-tracked conditions. Under the tracked condition, the equalization zone was steered to the location of the subject's head; under the untracked condition, the equalization zone remained at the ideal listening location.

- No scripted demonstration of the system was given. However, prior to each experimental session, 10 training trials were performed without feedback to refresh the subject with the protocol and the response method.

The individual experiments and results are described in the following sections.

5.2.1 Lateral head translations

These experiments tested localization using the loudspeaker system when the listener's head was translated 10 cm or 20 cm to the right of the ideal listening location. The source sound was the same set of pink noise pulses used in experiments 1 and 2. The binaural synthesizer and crosstalk canceller were the same as those used for experiment 2. The equalization zone was steered to the right by delaying and attenuating the right output channel as discussed in section 3.4.2. A 9 sample delay (0.28 msec at 32 kHz sampling rate) and an attenuation of 0.9 dB were used for the 10 cm translation, and an 18 sample delay (0.56 msec at 32 kHz sampling rate) and an attenuation of 1.8 dB was used for the 20 cm translation. It is doubtful whether the attenuations had much effect compared to the delays, but they were included for completeness.

3-D Audio Using Loudspeakers

Proper positioning of the subjects was ensured by appropriately positioning the visual sighting apparatus along the beam connecting the two loudspeakers. The apparatus was moved 10 cm to the right for the 10 cm head translations, and 20 cm to the right for the 20 cm head translations. As before, the subject's head position was measured relative to the speakers and the chair position was adjusted to obtain proper front-back positioning. With respect to the subject, the left and right speakers were at -36 and 23 degrees azimuth, respectively, for the 10 cm right translation, and -41 and 15 degrees azimuth, respectively, for the 20 cm right translation.

The same target locations were tested at each of the two head positions. A full circle of horizontal locations in 15 degree azimuth increments (24 locations) was tested with tracking enabled, and a full circle of horizontal locations in 30 degree increments (12 locations) was tested without tracking, for a total of 36 trials at each head position. These trials were randomly mixed. Half the subjects were tested first at 10 cm translation, and then at 20 cm translation, and the other subjects were tested in the opposite order. Subjects A, B, C, D, E, G, and H participated in these experiments.

Figure 5.11 shows the results as histograms of judged azimuth at each target azimuth for all seven subjects. For each of the two head positions, a histogram is shown for the untracked and tracked conditions. At the top are the results without tracking and at the bottom are the tracked results. At 10 cm translation without tracking (figure 5.11a) the localization results are poor. In the left hemisphere there is a great deal of variation in the responses, while in the right hemisphere it appears that subjects are primarily localizing the speaker at 23 degrees azimuth and the corresponding reversed azimuth at 157 degrees, regardless of the target azimuth. The results with tracking (figure 5.11c) are better, although there is still some asymmetry in the responses and many front-back reversals. On the left side, there is a tendency to localize towards 90 degrees, whereas on the right side there is a lot of variation for targets near 90 degrees. This makes sense, because listener movements to the right cause the left speaker to become more lateral and the right speaker to become more frontal.

The results for 20 cm translation are shown in figure 5.11b and figure 5.11d. These results have the same features as the 10 cm results, but now they are more exaggerated. The untracked results in figure 5.11b show the same tendency to localize the speaker positions (at -41 and 15 degrees) regardless of target azimuth. The tracked results in figure 5.11d are considerably better. Localization performance is better on the left side than on the right, where there is difficulty localizing extreme lateral targets.

Figure 5.12 shows the same data as figure 5.11 plotted as mean lateral angles with standard deviation errorbars. Only azimuths common to both tracked and untracked conditions, i.e. at 30 degree increments, are shown. The dashed line is the mean contour from the corresponding untracked results, shown to facilitate comparisons. It is

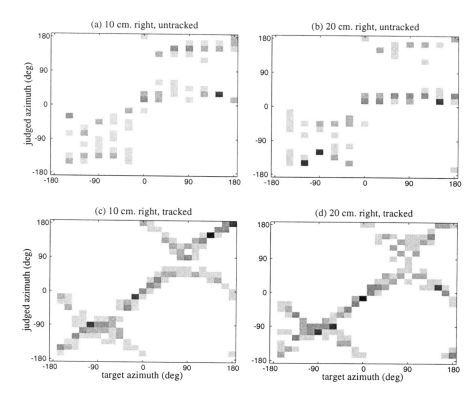

Figure 5.11 Histograms of judged azimuth at each target azimuth for the lateral head translation experiments: 10 cm. right translation, untracked (a), tracked (c), 20 cm. right translation, untracked (b), tracked (d). White indicates no responses, black indicates 100% response frequency. Under untracked conditions, righthand targets are localized primarily at the righthand speaker or its rear mirror location. Tracking greatly improves localization.

clear that tracking is increasing the angle of maximum lateralization for both 10 cm and 20 cm lateral translations.

The points marked with open diamonds indicate mean values which are significantly different under the two conditions, determined using a two-tailed matched pairs t test at a 5% significance level (Howell, 1997). For these target locations there is no more than a 5% chance that the tracked and untracked results were obtained by sampling distributions with the same mean. Many of the lateral target locations show statistically significant differences under the two conditions. This is true for both 10 cm and 20 cm lateral head translations.

The untracked results in figure 5.11 agree closely with similar results from Damaske (1971) and confirm the fact that localization performance degrades considerably when the listener is laterally translated away from the equalization zone. Damaske

3-D Audio Using Loudspeakers

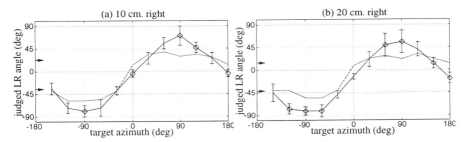

Figure 5.12 Mean judged lateral angle under tracked condition (solid lines with ±1 standard deviation errorbars) for 10 cm right head translation (a) and 20 cm right head translation (b), allsubjects. Dashed lines show mean values for untracked condition. Arrowheads on left show speaker positions. Points marked with open diamond indicate values that are significantly different ($p < 0.05$) under the two conditions. In both cases, the untracked results are similar to what we would expect from a conventional stereo pan, and tracking dramatically improves localization performance.

Table 5.2 Average angle error, front-to-back (F→B) and back-to-front (B→F) reversal percentages for lateral head translation experiments. Pairs of values are (untracked, tracked) conditions. Experiment 2 results are shown for comparison.

	Avg. angle error	F→B	B→F
10 cm right	(23.9°, 11.5°)	(37.1%, 37.7%)	(45.7%, 28.6%)
20 cm right	(27.6°, 16.4°)	(37.1%, 33.8%)	(57.1%, 39.0%)
experiment 2	12.1°	28.4%	36.4%

reported that a 10 cm lateral translation was sufficient to reduce the performance of the audio system to that of conventional two channel stereophony, which agrees with our findings. Sakamoto et al. (1982) reported that the maximum allowable lateral deviation for localization of male speech is 15 cm.

Table 5.2 summarizes the error statistics for the lateral head translation experiments. In both experiments, the average angle errors are much smaller for the tracked condition. In fact, the error for tracked 10 cm right translation, 11.5°, is smaller than the horizontal error in experiment 2, 12.1°. As explained in the chapter introduction, the initial experiments tested a full sphere of locations, and both azimuth and elevation judgements were gathered. We expect larger errors when the range of the stimuli and responses is increased; the errors are smaller in the head translation experiments because the task is restricted to horizontal localization. For purposes of comparison, it would have been more useful to conduct separate baseline experiments that tested only horizontal localization.

Front-to-back reversal percentages did not change much under the two conditions for either head translation. However, it appears as though tracking lessens the frequency of back-to-front reversals.

5.2.2 Front-back head translations

These experiments tested localization using the loudspeaker system when the listener's head was translated 16 cm to the front or 25 cm to the rear of the ideal listening location. The source was the same set of pink noise pulses used in the initial experiments. The binaural synthesizer and crosstalk canceller were implemented using the circuit shown in figure 3.30 on page 71. Binaural synthesis was accomplished by filtering the source with KEMAR HRTFs, sampled at 32 kHz. The HRTFs were free-field equalized with respect to 30 degree incidence, and implemented using 128-point FIR filters. The resulting binaural signal was processed by the shuffler crosstalk canceller. As shown in figure 3.30, the shuffler filters were comb filters containing a variable delay and a lowpass head shadowing filter. The lowpass head shadowing filter was implemented using an 8th-order IIR filter, designed by applying Prony's method to the lowpass filtered ITF shown in figure 15, chapter 3, which is based on KEMAR HRTFs at 30 degree incidence. The modeling delay m of the lowpass head shadowing filters was 4 samples (0.125 msec). The variable delays were implemented using first-order linear interpolation, although for this experiment the delays were rounded to integer numbers of samples.

For the 16 cm front translation, the speakers are at ±37 degrees azimuth with respect to the subject, corresponding to an ITD of 0.317 msec (10.1 samples at 32 kHz) for a spherical head model with diameter 17.5 cm (see equation 3.1 on page 31). For the 25 cm rear translation, the speakers are at ±23 degrees azimuth, corresponding to an ITD of 0.201 msec (6.4 samples at 32 kHz). The equalization zone was steered 16 cm to the front by increasing the ITD parameter to 10 samples (nominally 8 at the ideal listening position) or steered to the rear 25 cm by decreasing the ITD parameter to 6 samples.

Proper positioning of the subjects was ensured by using the visual sighting apparatus. Because the sighting apparatus casts two diverging beams of light, the angle between the beams needed to be changed for each of the two listening positions in order to obtain the correct interoccular spacing of the light beams at the listener position; this adjustment was accomplished by changing the distance between the point light source and the mask. As with the other experiments, the subject's head position was measured relative to the speakers and the chair position was adjusted to obtain proper front-back positioning.

The same target locations were tested at each of the two head positions. A full circle of horizontal locations in 15 degree azimuth increments (24 locations) was tested with tracking enabled, and a full circle of horizontal locations in 30 degree increments (12 locations) was tested without tracking, for a total of 36 trials at each head position.

3-D Audio Using Loudspeakers

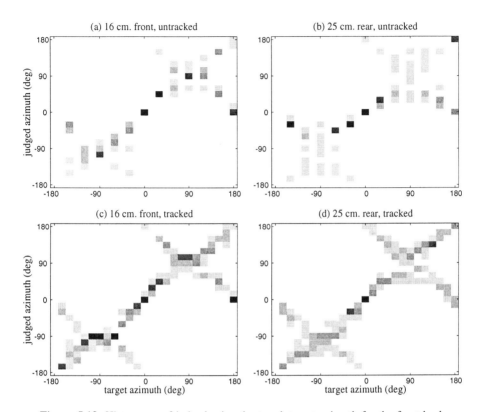

Figure 5.13 Histograms of judged azimuth at each target azimuth for the front-back head translation experiments, all subjects: 16 cm front translation, untracked (a), tracked (c), 25 cm rear translation, untracked (b), tracked (d). White indicates no responses, black indicates 100% response frequency. The results under untracked and tracked conditions are not obviously different. For the 25 cm rear translation, tracking seems to increase the slope of frontal responses and reduces the response variation at extreme lateral targets.

These trials were randomly mixed. Half the subjects were tested first at the front 16 cm translation, and then at rear 25 cm translation, and the other subjects were tested in the opposite order. Subjects A, B, C, D, F, G, and H participated in these experiments.

The results for front-back head translations are shown in figure 5.13. Figure 5.13a shows the results for the 16 cm forward translation without tracking and figure 5.13c shows the same results when the crosstalk canceller is adjusted to move the equalization zone forward. The results are not obviously different; with the exception of the ubiquitous front-back reversals, localization performance seems fairly good under both tracked and untracked conditions. We do see that the forward position of the listener causes a steeper slope of the responses near 0 degrees azimuth, expected

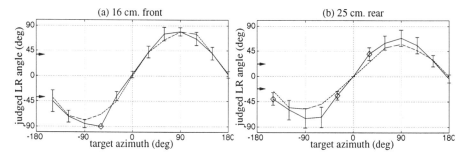

Figure 5.14 Mean judged lateral angle under tracked condition (solid lines with ±1 standard deviation errorbars) for 10 cm right head translation (a) and 20 cm right head translation (b), all subjects. Dashed lines show mean values for untracked condition. Arrowheads on left show speaker positions. Points marked with open diamond indicate values that are significantly different ($p < 0.05$) under the two conditions. In contrast to lateral head translations, front-back head translations do not cause localization performance to degrade considerably under untracked conditions. Some degradation occurs for the 25 cm rear translation that is apparently corrected by tracking.

because the speakers are now located at ±38 degrees azimuth, rather than at ±30 degrees. This is accompanied by a flattening of the responses at lateral positions.

Figure 5.13b and figure 5.13d show the same results for 25 cm rear translation. Though the results are similar, it appears as though the tracking is now improving performance. The rear listener position (the speakers are at ±23 degrees) tends to bias untracked frontal targets towards 0, resulting in a shallow slope of responses near 0, as seen in figure 5.13b. The ITD adjustment made in the crosstalk canceller to steer the equalization zone compensates for this bias, and indeed we see that the slope of the responses is indeed steeper in figure 5.13d. We did not see a similar slope correction with the front head translation. It also appears that the tracking is improving localization of lateral targets.

Figure 5.14 shows the same results plotted as mean lateral angles. For the 16 cm frontal translation, there appears to be little difference between tracked and untracked conditions. Only the -60 degree location is statistically different and it appears to be more in error than the untracked result. For the 25 cm rear translation it appears that the angles of maximum lateralization have increased in the tracked case, but these locations did not meet the 5% significance threshold. The use of more test subjects would make the statistical test more powerful, and we believe it would show that tracking has a statistically significant effect at lateral locations. The slope correction for frontal targets was seen to be statistically significant.

Table 5.3 summarizes the error statistics for the front-back head translation experiments. For the 16 cm front translation, the angle errors are very small for both untracked and tracked conditions. Tracking has more of an effect for the 25 cm rear translation, where localization performance begins to degrade slightly under the

3-D Audio Using Loudspeakers

Table 5.3 Average angle error, front-to-back (F→B), and back-to-front (B→F) reversal percentages for front-back head translation experiments. Pairs of values are (untracked, tracked) conditions. Experiment 2 results are shown for comparison.

	Avg. angle error	F→B	B→F
16 cm front	(10.7°, 9.9°)	(17.1%, 31.2%)	(42.9%, 28.6%)
25 cm rear	(15.2°, 10.6°)	(22.9%, 33.8%)	(60.0%, 39.0%)
experiment 2	12.1°	28.4%	36.4%

untracked condition. In both experiments, tracking increases front-to-back reversals, and decreases back-to-front reversals. Relative to the reversal rates in experiment 2, the untracked reversal rates in these experiments indicate there is a frontal bias of the untracked results. In other words, moving forward or backwards from the equalization zone increases the percentage of frontal responses under untracked conditions, a phenomenon that we can't explain. Tracking seems to correct this frontal bias, yielding reversal rates similar to experiment 2. The biggest improvement is seen for back-to-front reversals in the 25 cm rear translation, where tracking decreased the reversal rate from 60.0% to 39.0%.

5.2.3 Head rotations

These experiments tested localization using the loudspeaker system when the listener's head was rotated 20 degrees or 40 degrees to the left. At 40 degree left rotation, both speakers were on the righthand side of the subject's head. We expected difficulty in synthesizing lefthand images in this case, especially because high-frequency ILD cues would always suggest a righthand image.

Two sources were used: the pink noise pulses used in the initial experiments, and lowpass pink noise pulses obtained by lowpass filtering the pink noise pulses at a 6 kHz cutoff frequency. Because the crosstalk canceller is bandlimited to 6 kHz, the lowpass pulses only contain frequencies that are processed by the crosstalk canceller. We might expect better localization performance using the lowpass filtered pulses than using the unfiltered pulses, particularly for the case of -40 degrees rotation.

The binaural synthesizer and crosstalk canceller were implemented using the feedforward circuit shown in figure 3.29 on page 68, with a 6 kHz crosstalk cancellation cutoff frequency. The binaural synthesis included the high-frequency shelving filters discussed in section 3.4.7. The synthesis HRTFs, high-frequency shelving filters, and bandlimited crosstalk cancellation filters were all combined into a pair of 128-point FIR filters (at 32 kHz). A pair of filters was computed for each target location at each of the two head rotations. The source was filtered with the appropriate filter pair and the result was presented to the listener over loudspeakers.

Head orientation was controlled using the sighting apparatus described earlier. For the 20 degree left head rotation, the apparatus was positioned at 20 degrees left azimuth and angled to point at the ideal listening location. For the 40 degree rotation, the apparatus was positioned at 40 degrees left azimuth (to the left of the left louspeaker) and angled to point at the ideal listening location. The distance from the point light source to the mask was also adjusted to maintain the correct interoccular spacing at the listener position. As with the other experiments, the subject's head position was measured relative to the speakers and the chair position was adjusted to obtain proper front-back positioning. Subjects were instructed to report azimuths with respect to the rotated head position.

The same target locations were tested at each of the two head rotations. A full circle of horizontal locations in 30 degree azimuth increments (12 locations) was tested under both conditions of tracking enabled and disabled. This was done for both the pink noise pulses, and for the lowpass filtered pink noise pulses. Thus, at each head rotation, a total of 48 trials were performed in random order. Half the subjects were tested first at the left 20 degree rotation, and then at the 40 degree left rotation, and the other subjects were tested in the opposite order. Subjects A, B, C, D, E, F, G, and H participated in these experiments.

Figure 5.15a and figure 5.15c show the results for 20 degree left head rotation, pink noise source, untracked and tracked conditions, respectively. Note that both the target locations and the responses are relative to the rotated head orientation. It is apparent that localization performance is better on the righthand side under both tracked and untracked conditions. This is expected because the righthand speaker is at azimuth 50 degrees with respect to the listener, whereas the lefthand speaker is almost directly in front at azimuth -10 degrees. From the histograms it is difficult to see much difference between the tracked and untracked conditions. However, there is a righthand bias in the untracked results that is corrected in the tracked results. The bias towards the right is directly caused by the head rotation to the left. The tracked results also show better lateralization of lefthand targets than the untracked results.

Figure 5.15c and figure 5.15d show the results for 40 degree left head rotation, pink noise source, untracked and tracked conditions, respectively. In this case, both speakers are on the righthand side of the listener, and we expect great difficulty in synthesizing lefthand images. Indeed, the untracked results in figure 5.15d show very few responses in the left hemisphere. Tracking greatly increases the number of lefthand judgements, although there is considerable response variation. As with the 20 degree rotation results, we see an overall righthand bias in the untracked results that is corrected in the tracked results. Also, tracking improves localization performance for rear targets at 165 and 180 degrees azimuth.

3-D Audio Using Loudspeakers

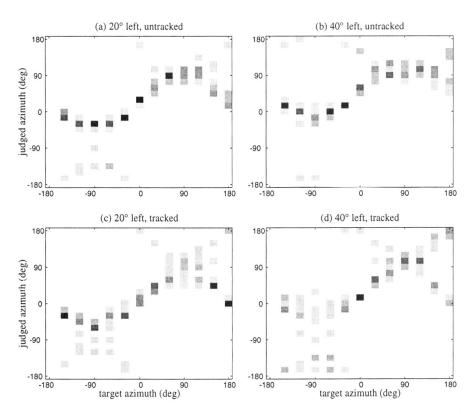

Figure 5.15 Histograms of judged azimuth at each target azimuth for the rotated head experiments using the pink noise source, all subjects: 20 degree left rotation, untracked (a), tracked (c), 40 degree left rotation, untracked (b), tracked (d). White indicates no responses, black indicates 100% response frequency. In both experiments, tracking increases the number of responses on the far left side, and corrects for a righthand response bias.

Figure 5.16 shows the same results in terms of mean judged lateral angle. In this form it is easy to see that tracking is correcting for the righthand bias in the responses caused by the left rotation of the listener's head. It is not possible to correct for this bias by simply rotating the target locations to match the head orientation. This strategy would decrease errors only for frontal targets. The results also show that tracking increases the angle of maximum lateralization for lefthand targets. Significant differences ($p < 0.05$) between the two conditions were found for many of the target locations.

Table 5.4 summarizes the error statistics for the head rotation experiments. In both experiments tracking greatly reduces the average angle error. For 40 degree left rotation, the average error under tracked condition is fairly large due to poor localization of lefthand targets. In both experiments, tracking had negligible impact on front-back

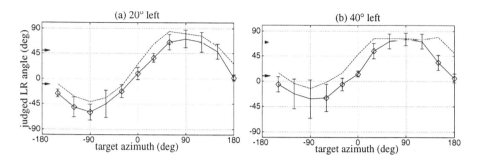

Figure 5.16 Mean judged lateral angle under tracked condition (solid lines with ±1 standard deviation errorbars) for 20 degree left head rotation (a) and 40 degree left head rotation (b). Dashed lines show mean values for untracked condition. Arrowheads on left show speaker positions. Points marked with open diamond indicate values that are significantly different (p < 0.05) under the two conditions. In both experiments, tracking greatly increases localization performance. For 40° left head rotation, tracked condition, lefthand targets yield mean lateral judgements on the left side of the head, despite the fact that both speakers are on the right side of the head. Nevertheless, it is difficult to synthesize extreme lateral left images in this case and there is considerable response variation.

Table 5.4 Average angle error, front-to-back (F→B) and back-to-front (B→F) reversal percentages for head rotation experiments, using pink noise source. Pairs of values are (untracked, tracked) conditions. Experiment 2 results are shown for comparison.

	Avg. angle error	F→B	B→F
20° left	(26.1°, 14.9°)	(15.0%, 17.5%)	(67.5%, 70.0%)
40° left	(43.7°, 25.3°)	(27.5%, 30.0%)	(47.5%, 50.0%)
experiment 2	12.1°	28.4%	36.4%

reversals. We note that for 20 degree left rotation, it appears that there is a bias towards frontal responses, relative to the results from experiment 2.

The results using the 6 kHz lowpass filtered pink noise pulses are nearly identical to the results using the unfiltered pink noise pulses, and we will not report them in detail. One notable difference was seen for 40 degree left rotation. Using the lowpass filtered noise reduced the response variation for lefthand targets, but did not increase the absolute mean lateral angle, relative to the results using the unfiltered noise pulses shown in figure 5.16.

It may seem surprising that it is possible to generate lefthand images when both speakers on the right side of the head. At low frequencies, crosstalk cancellation can effectively deliver ITD cues corresponding to a lefthand target even when both speakers are on the righthand side of the listener's head. Because crosstalk cancellation is limited to low frequencies, the high-frequency ILD cues will always suggest a righthand side target. However, the conflicting cues are always resolved in favor of the low-frequency ITD cues, provided the stimulus contains low frequencies (Wightman and Kistler, 1992). Thus, the auditory images are perceived on the left side of the head. However, it is not only difficult to synthesize extreme lateral lefthand images, but also there is a lot of response variation across subjects.. We note that the results using the 6 kHz lowpass filtered source had less response variation for lefthand targets. This would support the notion that the presence of conflicting high-frequency cues increases response variation. It is also possible that intersubject variation in ITFs may be greater for large incidence angles, which would increase intersubject variation in localization performance under these conditions

Despite the conflicting spatial cues, the auditory image remains fused and localized to a compact spatial region. The cues for fusion of the auditory image (Bregman, 1990) must dominate the spatial cues, otherwise we would expect two auditory images: a left side image consisting of low frequencies, and a right side image consisting of high frequencies. We attribute the auditory fusion to the strong common onset cue present in the pink noise pulses.

5.3 DYNAMIC HEAD MOTION

This section describes an experiment that tested localization of loudspeaker synthesized images during head motion. The purpose of the experiment was to validate the hypothesis that tracking the head position and appropriately adjusting the binaural synthesis and crosstalk cancellation can be used to deliver useful dynamic localization cues to a listener. The importance of dynamic localization cues was demonstrated by Wallach (1939, 1940) using an array of loudspeakers to present the stimulus to the listener. The importance of dynamic localization cues has also been validated for headphone presentation of binaural audio (Boerger et al., 1977; Wenzel, 1995; Wightman and Kistler, 1997). Our experiment is conceptually similar to Wallach's, but we have replaced the array of loudspeakers with two loudspeakers that deliver binaural audio.

The binaural synthesizer and crosstalk canceller were essentially the same as used for the static head rotation experiments, except that we allowed arbitrary horizontal translations and rotations of the listener's head. The listener's head position relative to the speakers was parameterized in terms of the angular spacing of the speakers, the rotation of the head with respect to the midpoint between the speakers, and the distance from the head to each of the speakers. A pair of 128-point FIR filters containing the synthesis HRTFs, the crosstalk canceller, and the high-frequency shelving filters was computed for each combination of synthesis location, head rotation, and speaker width angle. Any difference in path length to the two speakers was compensated by appropriately delaying and attenuating the closer speaker, as described in section 3.4.2. Synthesis locations were chosen on the horizontal plane in increments of 5 degrees azimuth, absolute head rotations were chosen from 0 to 45 degrees in increments of 5 degrees, and speaker width angles were chosen from 55 to 70 degrees in increments of 5 degrees. The set of all combinations comprised 2880 filter pairs. Separately implementing the binaural synthesis and crosstalk cancellation filters would greatly reduce filter coefficient storage requirements, but this was not a concern for experimentation purposes.

Head tracking was accomplished using a Polhemus[†] ISOTRAK motion tracker that was worn on the head of the listener. Position updates were sent to the computer every 25 msec. The computer processed sound in blocks of 128 samples at a sampling rate of 32 kHz (one block every 4 msec). Each block of input sound was filtered with the two FIR synthesis filters using FFT based block convolution. At each head position update (every 6 or 7 blocks), the listening geometry was computed and a new pair of synthesis filters was read from a table. The input sound was filtered with both the old filter pair and the new filter pair and the output was obtained by a linear crossfade between the old and new outputs. The crossfade prevented head motion from creating clicks in the output sound. The total latency from head motion to an audible change at the listener's ear was calculated to be 88 msec maximum, but this was not confirmed by measurement. Much of this latency was due to the 40 msec output audio buffer in the computer.

The source sound and experimental protocol for this experiment differed from the earlier experiments. The source was a single 250 msec pink noise pulse with 10 msec linear onset and offset ramps. In this experiment, the subject initiated the stimulus presentation by rotating his/her head. Subjects were instructed to face the left loudspeaker and to rotate their head to face the right loudspeaker. The subject's head rotation automatically triggered the stimulus presentation when the head was at -10 degrees azimuth. Triggering was conditional on the rotation rate; rates from 50 degrees/sec to 200 degrees/sec triggered the stimulus. A typical rate of 80 degrees/sec caused the 250 msec stimulus to be emitted while the subject subtended -10 to +10 degrees azimuth. Statistics on head rotation rates were not recorded.

Prior to each experimental session, the subject was seated, the head tracker was donned, and the subject was positioned in the ideal listening location. The subject was asked to view the sighting apparatus in order to ensure that the subject's head was in the ideal listening location. At this time, the experimenter pressed a button that initiated an automatic calibration of the head tracking apparatus that corrects for the orientation of the tracking sensor on the subject's head. The sighting apparatus was then turned off for the remainder of the experiment. The subject was then instructed in the head rotation task, and allowed to practice. Then a set of 10 training trials without feedback was performed. Prior to each trial, the experimenter announced "OK", the subject then performed the rotation task which triggered the stimulus, and then the

---

[†]Polhemus, P.O. Box 560, Colchester, VT 05446.

3-D Audio Using Loudspeakers

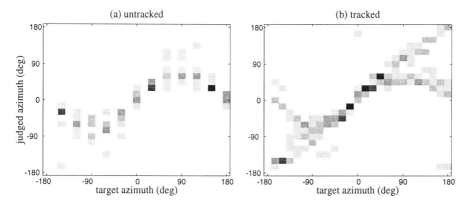

Figure 5.17 Histograms of judged azimuth at each target azimuth for dynamic head rotation experiment, under untracked (a) and tracked (b) conditions, all subjects. All targets are on the horizontal plane. White indicates no responses, black indicates 100% response frequency. In the untracked condition almost all targets are perceived in the front. Tracking greatly decreases the number of back-to-front reversals.

subject verbally reported the perceived location of the sound, giving azimuth and distance judgements only.

A full circle of horizontal locations in 15 degree azimuth increments (24 locations) was tested under the condition of tracking enabled, and a full circle of horizontal locations in 30 degree azimuth increments (12 locations) was tested under the condition of tracking disabled Trials were presented in random order. Under the untracked condition, head motion did not affect the sound processing, but the head tracker was of course still functioning to allow triggering of the stimulus. Subjects were instructed to report the location of perceived sound with respect to frontal orientation, and to report the midpoint of a sound trajectory if the sound appeared to move during presentation. Subjects B, C, D, E, F, G, and H participated in this experiment.

Results for the dynamic head motion experiment are shown in figure 5.17, which shows histograms of judged azimuth at each target azimuth for tracked and untracked conditions. It is clear that tracking is principally affecting front-back reversals. Under the untracked condition (figure 5.17a), almost all targets are perceived to be frontal. Under the tracked condition (figure 5.17b), frontal targets are still correctly perceived as frontal, and many of the rear targets are now correctly perceived in the rear. The reversal percentages are given in table 5.5. For the untracked condition, 91.4% of rear targets are reversed to the front, and only 8.6% of frontal targets are reversed to the rear. For the tracked condition, the back-to-front reversal rate decreased to 50.7%, and the front-to-back reversal rate remained low at 6.5%. The reversal rates are plotted for each subject in figure 5.18. Under untracked conditions, subjects C, D, E, F, G, and H reversed all rear targets to the front. Under tracked conditions, the back-to-front reversal rates decreased considerably for subjects C and D,

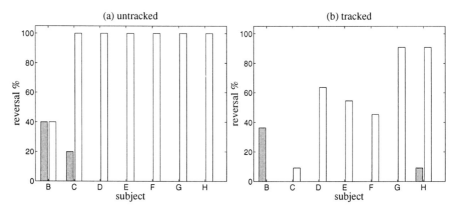

Figure 5.18 Bargraphs of individual front-to-back (gray bars) and back-to-front (white bars) reversal percentages for dynamic head rotation experiment, under untracked (a) and tracked (b) conditions. Back-to-front reversals decreased greatly for subjects B-F under the tracked condition.

Table 5.5 Average angle error, front-to-back (F→B), and back-to-front (B→F) reversal percentages for the dynamic head rotation experiment. Pairs of values are (untracked, tracked) conditions. Experiment 2 results are shown for comparison.

	Avg. angle error	**F→B**	**B→F**
dyn. rotation	(14.8°, 14.4°)	(8.6%, 6.5%)	(91.4%, 50.7%)
experiment 2	12.1°	28.4%	36.4%

decreased by about half for subjects E, F, and G, and decreased only slightly for subjects G and H.

Figure 5.19 plots the results in terms of mean judged lateral angle across all subjects. Unlike the previous fixed head experiments, the results under the two conditions were not found to be significantly different ($p < 0.05$) at any target location. This is not completely unexpected, because the listeners' heads were almost exactly in the ideal listening location when the stimulus was presented, and the stimulus was of short duration.

In Wallach's experiments, subjects required a brief period of head motion before the dynamic cues established themselves (Wallach, 1939, 1940). In our experiment, the stimulus was only 250 msec, and yet it is clear the dynamic cues had a considerable effect on front-back reversals. It is likely that more improvement in back-to-front reversals would result if the subjects were allowed to move their heads freely while localizing a long stimulus. We opted for a more constrained experiment so that the

3-D Audio Using Loudspeakers

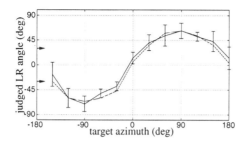

Figure 5.19 Mean judged lateral angle as a function of target azimuth for dynamic head rotation experiment under tracked condition (solid line with ±1 standard deviation errorbars) and untracked condition (dashed line), all subjects. Arrowheads on left show speaker locations.

listener's head was always within a region in which the tracked crosstalk cancellation was effective.

5.4 DISCUSSION

The initial experiments comparing localization of virtual sources over headphones and loudspeakers showed important differences between the two presentation methods. Headphones had great difficulty reproducing frontal images using non-individualized HRTFs. The KEMAR HRTFs seem particular poor in this respect, but we note that 50% of frontal locations were reversed to the rear in Wenzel's study, using the HRTFs of a good human localizer (Wenzel et al., 1993). Furthermore, over headphones, images tended to be localized very close to the head, particularly for medial targets. In contrast, loudspeakers imaged the frontal locations quite well, with good externalization. Given that frontal locations are of particular importance for multimedia applications, loudspeakers would seem to be a superior choice to headphones for non-individualized implementations. Presenting sounds over frontally placed loudspeakers incorporates the listener's individualized frontal spectral cues into the sound, which should bias responses to the frontal direction. Indeed, the experimental results over loudspeakers tended to be biased to the front. This phenomenon has resulted in headphone designs that use transducers placed frontally away from the listener's ears, clearly intended to improve frontal localization.

Headphones do have advantages compared to frontally placed loudspeakers, namely superior rear imaging and elevation localization. Headphones are particularly effective when individualized HRTFs are used, where localization performance is essentially unchanged from free-field conditions (Wightman and Kistler, 1989b; Moller et al., 1996a). It is of course possible to construct a loudspeaker system using individualized HRTFs for both binaural synthesis and crosstalk cancellation. Such a system has been built by Koring and Schmitz (1993) and the performance results of their system in both anechoic and reverberant conditions were better than our non-individualized system based on KEMAR HRTFs. We would not expect an individualized loudspeaker system to perform as well as an individualized headphone system because of interference due to high-frequency crosstalk and reverberation.

In all the experiments that compared tracked and untracked conditions, tracking was seen to increase localization performance in terms of average angle error. For lateral head translations and head rotations, this improvement was dramatic. A symmetric crosstalk canceller combined with a variable delay on one of the output channels can effectively steer the equalization zone to a laterally translated listener. This is despite the fact that the listening situation is asymmetric. For instance, the 20 cm right head translation is roughly equivalent to a 13 degree right head rotation, which is just beyond the 10 degrees we have previously described as the maximum allowable (untracked) head rotation. Nevertheless, the tracked performance in this case was quite good. However, better performance might well result if an asymmetric implementation was used, or if the right head translation was accompanied by a 13 degree left head rotation.

The symmetric crosstalk canceller cannot in general compensate for a rotated head; it is necessary to use an asymmetric crosstalk canceller. Furthermore, if the crosstalk canceller is bandlimited, a high-frequency power compensation circuit should also be used. The experiments demonstrated the importance of compensating for a rotated head. However, the experiments did not specifically test the importance of the high-frequency power compensation; to do this would require comparing localization both with and without the power compensation circuit.

The rotated head experiments show the difficulty of synthesizing extreme lateral sources on one side of the listener when both loudspeakers are on the opposite side of the listener's head. There doesn't appear to be an easy solution to this problem, except to use wider spaced speakers, or to use more speakers, solutions that may be impractical for desktop applications. Fortunately, in a typical desktop application, we would expect the listener to have his/her gaze directed towards a video monitor placed between the two loudspeakers, in which case the speakers will likely be on opposite sides of the listener's head.

The final experiment demonstrated that a head-tracked virtual acoustic display using loudspeakers can provide useful dynamic localization cues when the listener's head is rotating. These cues considerably decrease front-back reversals. Based on Wallach's results (Wallach, 1939, 1940), we might expect that these dynamic cues can also help in elevation localization, but this remains to be tested. Wenzel (1995) has shown that ILD cues are more important for dynamic localization than ITD cues. The bandlimited crosstalk canceller is more effective at delivering proper low-frequency ITD cues than high-frequency ILD cues. On this basis we would expect the improvement of adding dynamic cues to be greater when using headphones than when using loudspeakers.

6 DISCUSSION

6.1 CONTRIBUTIONS OF THIS WORK

The central thesis of this work is that loudspeaker binaural audio can be sent to a moving, tracked listener by dynamically inverting the acoustic transmission paths, and that doing so greatly improves localization performance relative to the untracked condition. Localization performance improves both because the equalization zone is steered to the position of the tracked listener, and also because dynamic localization cues are enabled. We have proposed, studied, and validated these concepts. Furthermore, we have constructed a working implementation of a head-tracked 3-D loudspeaker audio system. As part of this process, we have made a number of subsidiary contributions to the field, listed below.

An extensive set of KEMAR HRTFs measurements was made, and the data are publicly available on the Internet (Gardner and Martin, 1994, 1995). A comparison of our localization study and that of Wenzel et al. (1993) has shown that the KEMAR HRTFs are not as good for synthesizing spatial cues as the HRTFs of a good human localizer. Nevertheless, the dense spatial sampling of the KEMAR HRTF data make them a good candidate for various types of structural analysis, such as the study by Lopez-Poveda and Meddis (1996). The KEMAR HRTFs have also been shown to produce a good "typical" head shadowing model.

A theory of crosstalk cancellation was developed that is sufficiently general to describe both existing symmetric crosstalk cancellers and the steerable, asymmetric crosstalk cancellers proposed herein. Some of the proposed designs are based on an embedded head shadowing model, implemented as a causal filter $L(z)$ cascaded with some delay. Methods for deriving the head shadowing model from measured ITFs of human listeners and approximating these functions with low-order IIR filters have been described. The embedded head shadowing model has been used to create efficient, recursive implementations of both symmetric and asymmetric crosstalk cancellers. These filters have been shown to be stable and realizable for frontally facing listeners. Simplifications in the filter approximations can yield crosstalk canceller implementations that are very computationally efficient.

A hybrid strategy for crosstalk cancellation has been developed, consisting of a band-limited crosstalk canceller operating at low frequencies, and a power compensation circuit operating at high frequencies. The idea of bandlimiting the crosstalk cancellation is not new (Cooper and Bauck, 1990), but the high-frequency power model is new. The high-frequency power compensation strategy achieves only modest improvements for symmetrically positioned listeners, but is essential for delivering optimal high-frequency ILD cues to rotated listeners. Efficient methods for implementing the high-frequency power model have also been described.

Physical measurements and acoustical simulations have been undertaken to validate both crosstalk cancellation performance and also the concept of steering the equalization zone. Similar validation work has been performed by other authors; for instance, Kotorynski (1990) has validated crosstalk cancellation performance via simulation, and Nelson et al. (1995) and Asano et al. (1996), for example, have shown contour plots of simulated equalization zones. These techniques are indispensible for the study of crosstalk cancellation. Although it is not difficult to do, ours is the only study that has evaluated non-individualized crosstalk cancellation at the ears of human listeners.

Finally, extensive sound localization experiments have been conducted to validate these concepts from a psychoacoustical standpoint. Other authors have conducted sound localization experiments using loudspeaker binaural audio systems (Damaske, 1971; Sakamoto et al., 1982; Koring and Schmitz, 1993). Our experiments are unique in several respects. First, a comparative study was made of headphone and loudspeaker systems based on the same non-individualized synthesis HRTFs. The results of this study show quite clearly that the loudspeaker system is better at reproducing frontal and externalized images, but is poorer at reproducing elevated images. Second, our experiments studied the effectiveness of steering the equalization zone to a listener who is displaced from the ideal listening location, either via a lateral or front-back translation, or via a head rotation. The results demonstrated significant improvements in localization performance when the equalization zone is steered to the location of the listener, relative to the unsteered condition. Finally, the experiments tested dynamic localization using a head-tracked loudspeaker system. The results show a substantial decrease in front-back reversals when dynamic cues are enabled. We conclude that binaural audio can be delivered to a moving, tracked listener, and that doing so improves localization performance both because the equalization zone is correctly situated, and also because dynamic localization cues are enabled.

6.2 CHALLENGES TO GENERAL USE

Significant challenges prevent this technology from being used in general listening situations, such as a living room. Besides the head tracking issue, which we leave to others to solve, there are the audio related problems posed by multiple listeners and room reverberation. Here we elaborate on these issues.

6.2.1 Multiple listeners

The linear equations that describe the transmission path inversion do not preclude the possibility of multiple listeners. As previously discussed, many authors have proposed loudspeaker systems that can deliver binaural audio to multiple listeners by inverting the transmission matrix (Yanagida et al. 1983; Miyoshi and Kaneda, 1988; Bauck and Cooper, 1992, 1993, 1996; Abe et al., 1995; Nelson et al., 1995), but it is not known whether any of these systems have been implemented. Besides the implementation complexity issue, there are several problems associated with multiple listeners. First, a listening situation involving multiple listeners, such as a living room, will probably be larger than a single listener situation, and therefore the listeners will be relatively farther from the loudspeakers and subject to a lower direct-to-reverberant ratio of sound. Second, the transfer functions from a speaker to the ears of a listener may depend significantly on the positions of the other listeners, especially if the acoustical path from a speaker is interrupted by another listener. This adds enormous complexity to the modeling of the acoustic transfer matrix. Finally, there is the problem of high-frequency crosstalk cancellation. The single listener methods we have discussed work despite the lack of high-frequency cancellation, due to the combination of the dominance of low-frequency time cues and the naturally occuring high-frequency head shadowing. However, with multiple listeners we may desire to send a different audio program to each listener. Two possibilities are listed below:

> Each listener hears an entirely different audio program. This requires the complete acoustical cancellation of unintended program material at each listener, which is not possible without the use of individualized head models, extremely accurate head tracking, and room inverse filtering. Inverting the room response for a possibly moving listener is a particularly difficult problem. Room reverberation has a complex spatial dependence, it is somewhat time varying, and it depends on the positions of all listeners in the room. Although dynamic room inversion might be possible at low frequencies, we must accept significant high-frequency crosstalk between listeners.

> Each listener hears the same audio sources at possibly different spatial positions (relative to each listener). This situation naturally arises when a number of people are listening to the same auditory scene but are individually oriented differently. If the individually differing synthetic source positions can be rendered by manipulating the low-frequency cues, then the high-frequency crosstalk (both between and within listeners) may be acceptable; it is analogous to the single listener situation. Because all listeners are hearing the same sources, the room reverberation doesn't constitute "leakage" from unwanted sources; provided the reverberation doesn't substantially interfere with localization, it shouldn't be perceptually objectionable.

6.2.2 Distant listening

The use of near-field loudspeakers increases the ratio of direct to reverberant sound in the listening space. Listening space reverberation degrades the performance of the 3-D audio system in much the same way as reverberation affects natural localization; it degrades both interaural and monaural localization cues, and competes with the direct sound for the auditory system's attention.

We have not conducted experiments that separately assess the effects of reverberation on loudspeaker 3-D audio systems. All of our experiments were conducted using near-field loudspeakers (76 cm = 30 in distance) in a mildly reverberant room (RT = 240 msec at 500 Hz). Informal tests in more reverberant surroundings revealed significant localization degradation, particularly in the form of increased back-to-front reversals. Damaske (1971) and Sakamoto et al. (1982) showed that increasing reverberation caused increased frequency of back-to-front reversals. Koring and Schmitz (1993), using an individualized system, showed that reverberation slightly increased horizontal errors, and greatly increased both front-to-back reversals and elevation errors. Horizontal localization was quite good, even with a one second reverberation time and 3 meter speaker to head distance. Overall, these results are compatible with the notion that reverberation primarily degrades spectral cues; thus, front-back reversals and elevation localization are most seriously affected. This offers promise that head-tracked loudspeaker 3-D audio could improve horizontal localization accuracy for single listeners who are relatively distant from the loudspeakers..

In natural listening, the auditory system makes use of various *precedence* mechanisms to suppress the distorting effects of reverberation (Rakerd and Hartmann, 1983; Hartmann, 1983, 1997; Zurek, 1987). Hartmann (1997) describes three different precedence effects, each operating over a different time span. The "law of the first wavefront" describes the auditory system's ability to determine the location of a sound from the initial onset; localization information in subsequent echoes is largely suppressed, even when the echoes are up to 10 dB more intense than the initial sound. This precedence mechanism is principally responsible for our ability to localize sounds in reverberant environments.

In the context of loudspeaker binaural audio, the precedence effect should in theory allow us to localize synthetic sounds in the presence of competing listening space reverberation. Provided the first wavefront is correctly rendered at the listener, the precedence mechanism should capture the proper localization information in the onset and suppress the subsequent reverberation, just as would happen with a real source.

The bandlimited crosstalk cancellers we have discussed do not in fact render a perfect first wavefront. The first wavefront will contain low-frequency cues that correspond to the virtual source, and high-frequency cues that correspond to the real sources (the loudspeakers). Thus, bandlimited crosstalk cancellers (in anechoic conditions) could be modeled as a low-frequency source at the target location, plus two coherent high-

3-D Audio Using Loudspeakers 137

frequency sources at the loudspeaker locations. For lateral targets, the high-frequency power compensation circuit shuts off the contralateral speaker, leaving only a single high-frequency source. It would be interesting to perform localization experiments using simultaneous low and high-frequency sources at different directions.

The study by Wightman and Kistler (1992) shows that low-frequency ITD cues dominate ILD and spectral cues. However, care must be taken in applying these results to bandlimited crosstalk cancellation, because the situations are not exactly the same. Wightman and Kistler's study used stimuli created by combining the time cues for one direction with the intensity cues for another direction. It was then observed that localization judgements substantially followed the ITD cues, but only when the stimuli contained low frequencies. The stimuli used are not the same as simultaneous low and high-frequency sources at different directions.

6.3 DIRECTIONS FOR FUTURE WORK

This study has uncovered many avenues for research in this area. Some specific ideas are listed here.

Parameterized head-shadowing. As discussed in section 3.1.1, crosstalk cancellers require a head model that is *acoustically* accurate. Much of the current work in parameterization of HRTFs is intended to yield *perceptually* valid models. Thus there is a need for a systematic study of human HRTFs, or more specifically, ITFs, in order to obtain an optimized non-individualized head shadowing model. This could take the form of a "typical" ITF, i.e., an individual or average head model that works well for most listeners. Our choice of the KEMAR, which is based on median human measurements, was intended to yield a reasonably typical head-shadowing model. Ideally, we would like to have a parameterized acoustical model that can be customized for a particular listener. The parameters could be set via a calibration task, or perhaps using head geometry data obtained by the head tracker. One important parameter is head size, or equivalently ITD, which as we have seen accounts for a great deal of variation in crosstalk cancellation performance. The head shadowing model we have developed (equation 3.7 on page 40) does contain a separate ITD parameter. There may be other geometrical or shape parameters that account for other variations in ITFs across subjects. A principal component analysis of low-frequency ITFs across multiple subjects might be a fruitful place to begin.

Individualized head models. Our head-tracked 3-D audio system would obviously work better using individualized head models. In fact, an individualized model would allow the possibility of high-frequency crosstalk cancellation, provided the head tracking was sufficiently accurate. Although individualized head models have been used in crosstalk cancellers (Koring and Schmitz, 1993), they haven't yet been tried with a dynamic, head-tracked system.

Loudspeaker placement. Kulkarni has suggested that crosstalk canceller speakers should be mounted on opposite sides of the head at high elevations, so that the speaker to ear transfer functions are relatively flat[†]. In fact, an examination of figure 3.4 on page 33 reveals that the KEMAR HRTFs do not contain sharp features at elevations above 60 degrees. This seems to agree with human HRTF data measured by Moller et al. (1995c). If indeed this is a general property of HRTFs, then placing speakers at high elevations avoids the problem of inverting notches in the ipsilateral response, which creates objectionable peaks in the inverse response.

Kulkarni has also demonstrated that an overhead loudspeaker emiiting noise bursts can be used to generate auditory images at lower elevations when a notch filter is applied in the 5-10 kHz range; the elevation of the image is directly related to the frequency of the notch, as suggested by the N1 feature in figure 3.4. It remains to be seen whether this is a viable arrangement for a spatial auditory display. We have already discussed the possibility that frontally placed loudspeakers greatly assist the successful perception of externalized frontal imagery. Overhead loudspeakers may not be able to convicingly reproduce frontal imagery.

Closely spaced, frontal loudspeakers on the horizontal plane are well suited to equipment that must be spatially compact, e.g., a small monitor, laptop computer, portable radio, etc. We have discussed the possibility that closely spaced loudspeakers can increase the size of the equalization zone. In fact, the simulation results in figure 4.5 show only modest improvement in the width of the equalization zone when using closely spaced loudspeakers. One disadvantage of using closely spaced loudspeakers is the relative lack of natural high-frequency channel separation due to head shadowing. Another problem particular to the head-tracked approach is that small head rotations will cause both loudspeakers to fall on the same side of the head. These issues could be studied further.

Structural averaging of HRTFs. As previously discussed, an ipsilateral inverse filter will exhibit a sharp peak wherever the ipsilateral response has a notch. Our approach has been to smooth the inverse ipsilateral response with a 1/3-octave averaging filter, as shown in figure 3.7 on page 36. The approach taken by Koring and Schmitz (1993) is to derive *both* the binaural synthesis filters and the crosstalk canceller filters from HRTF data that has been smoothed in this manner. We believe this has several advantages over our approach. First, the use of smoothed HRTFs should reduce the timbral aberrations that are currently noticed with our system during head and virtual source motion. These timbral problems are caused by sharp spectral features changing frequency. Second, the use of smoothed HRTFs for both binaural synthesis and crosstalk cancellation will cause the transfer function from source to loudspeaker to be exactly flat when the source is panned to a loudspeaker position,

---

[†]Abhijit Kulkarni, personal communication, 1996.

which is the power panning property. With our system, this transfer function contains the residual between the smoothed and unsmoothed ipsilateral response.

Room equalization. Room equalization techniques have not been used in our study. It is easy to imagine that the tracked position of a listener could be used to index a room equalization filter previously calculated for that listening location. It would be interesting to see whether room equalization techniques can improve the reproduction of binaural audio from loudspeakers, especially when the listener is relatively distant from the speakers. We would expect room equalization to improve spectral cues; therefore front-back and elevation localization would also improve.

Multiple channel implementations. A multiple channel (i.e., more than two) head-tracked system would be relatively easy to build. A four channel system could be constructed using speakers at the corners of a square. When the listener faces front or back, the speakers would be logically grouped into two crosstalk-cancelling pairs consisting of (front-left, front-right) and (rear-left, rear-right). When the listener faces to the left or right, the speakers would be grouped as (left-front, left-rear) and (right-front, right-rear). Such a system would allow unencumbered listener rotations, and would be equally effective at rendering sound in any direction on the horizontal plane. We expect such a crosstalk cancelling system to perform better than an equivalent discrete multi-channel system that uses intensity panning. Do the respective performances converge as the number of channels increases? Also, when the speakers are not all on the horizontal plane, how does one group the speakers into crosstalk cancelling pairs?

Efficient implementations. As discussed in section 3.4.2, extremely efficient implementations exist for the asymmetrical crosstalk canceller shown in figure 3.27 on page 65; the head shadowing filters can be implemented with first-order lowpass filters, the variable delay lines can be implemented with linear interpolators, and the ipsilateral equalization filters can be omitted entirely. These simplifications should be tested.

High-frequency power model. The high-frequency model described in section 3.4.6 could be improved. The analysis of high-frequency power transmission could be extended to multiple frequency bands, leading to a multiple band shelving filter implementation. This would be justified if sufficient similarity of band-averaged, high-frequency HRTFs was found between different subjects, and if the individual band behaviors were sufficiently different to warrant separate treatment. Ultimately, the division of the high frequencies into finer bands would lead to a general implementation of a continuous high-frequency power compensation filter.

Another improvement to the high-frequency power model would be to rethink the assumption that the crosstalk signals add incoherently at the ears. In the absence of crosstalk cancellation, each ear receives the sum of the intended ipsilateral signal and the crosstalk from the opposite ear. Assuming a single source is synthesized, the delay between the ipsilateral and crosstalk components depends on both the ITD of the listener's head shadowing and the ITD in the synthetic HRTFs. For the purposes of power calculation, if we assume broadband signals, we are justified in assuming incoherent addition because the delay between the components is likely to be much larger than the period of the high-frequency signals. However, the addition of the two delayed components causes a comb filtering effect that we suspect interferes with the monophonic spectral localization cues, especially for medial sources. It is unclear whether these comb filtering effects can be predicted and equalized, and whether this is in practice any different from extending crosstalk cancellation to higher frequencies. It should be noted that this phenomenon only occurs for near-medial sources, where both loudspeakers are emitting similar high-frequency powers.

Head tracking. Much work remains to be done integrating this technology with various head tracking modalities. We currently have plans to combine our system with the video based head-tracker of Oliver et al. (1997); this will free users from having to don the tracking apparatus. Indeed, a primary motivation of using loudspeakers is to avoid the requirement of wearing headphones. Our studies of the size of the equalization zone (in Chapter 4) and the degradation of localization performance with displaced listeners (in Chapter 6) suggest that the head tracker should have lateral accuracy within a few cm, and rotational accuracy within 5-10 degrees. Two diffferent visual head trackers developed at the Media Lab by Basu et al. (1996) and Oliver et al. (1997) both have lateral accuracy within 1 cm. The tracker by Basu et al. (1996) has orientation accuracy within 5 degrees. Thus, visual head trackers perform accurately enough for this application; the challenge is to make them robust and computationally efficient. It will also be interesting to see if estimation techniques can reliably predict the position of the listener's head based on recent data. If successful, this will allow relatively slow frame rate tracking technology to be used without any penalty of increased tracking latency.

A INVERTING FIR FILTERS

In this section we review methods to construct inverse filters for finite length, discrete time signals, such as head-related impulse responses. An example of a discrete-time signal with N samples is:

$$h[n], 0 \leq n < N \tag{A.1}$$

which has a z-transform of:

$$H(z) = h[0] + h[1]z^{-1} + h[2]z^{-2} + \ldots + h[N-1]z^{N-1} \tag{A.2}$$

Because $h[n]$ is a finite impulse response (FIR) signal, $H(z)$ can be fully described in terms of its roots, or zeros, in the z-plane (all of its poles are at $z = 0$). The zeros will either be minimum-phase (inside unit circle), maximum-phase (outside unit circle), or will fall on the unit circle. In this latter case, the inverse is undefined where the frequency response is zero, so we restrict our discussion to signals with non-zero frequency responses (i.e., no zeros on the unit circle). The inverse filter for $h[n]$ is the signal $g[n]$ such that $h[n] * g[n] = \delta[n]$, where * denotes convolution and $\delta[n]$ is the unit impulse signal. The z-transform of $g[n]$ is:

$$G(z) = \frac{1}{H(z)} \tag{A.3}$$

The inverse $G(z)$ will have poles where $H(z)$ has zeros, and will have all of its zeros at $z = 0$. In order to obtain a stable $g[n]$, we choose the region of convergence (ROC) of the z-transform to include the unit circle. Any poles inside the unit circle correspond to causal exponentials in the time domain, and poles outside the unit circle correspond to anti-causal exponentials. Thus, the stable inverse filter $g[n]$ will be infinitely long, and its time support (causal, anti-causal, or two-sided) depends on the distribution of zeros in $H(z)$.

When $h[0] \neq 0$, we can rewrite equation A.3 as:

$$G(z) = \frac{\dfrac{1}{h[0]}}{1 + \dfrac{h[1]}{h[0]}z^{-1} + \dfrac{h[2]}{h[0]}z^{-2} + \ldots + \dfrac{h[N-1]}{h[0]}z^{-(N-1)}} \quad (A.4)$$

which has an obvious recursive implementation. However, the resulting system will be stable if and only if $H(z)$ is minimum-phase.

When $H(z)$ is non-minimum-phase, the inverse filter will have infinite, two-sided time support, and the recursive filter structure suggested by equation A.4 will be unstable. We can obtain a finite-length inverse filter that approximates the true inverse by windowing the true inverse:

$$f[n] = g[n] \cdot w[n], \quad w[n] = \begin{cases} 1, & L \leq n < M \\ 0, & \text{otherwise} \end{cases} \quad (A.5)$$

For a given window length $M-L+1$, we choose L so that the energy of the signal $g[n]$ falling under the window is maximized. The use of a rectangular window, as specified, results in minimizing the squared error between $F(e^{j\omega})$ and $G(e^{j\omega})$. If $L < 0$, the filter $f[n]$ is not causal. A realtime implementation would use the causal filter $f[n+L]$, thus applying a modeling delay of $-L$ samples.

We now discuss how to determine the time response $g[n]$. One method is to explicitly evaluate the inverse z-transform of $G(z)$ to solve for $g[n]$ using partial fraction expansion or the Cauchy residue theorem (Oppenheim and Schafer, 1989). However, these techniques can be numerically unstable for certain inputs. A more practical method is to use the discrete Fourier transform (DFT). The N-point DFT of $h[n]$ is

$$H[k] = \text{DFT}\{h[n]\} = \sum_{k=0}^{N-1} h[n]e^{(-j2\pi kn)/N} = H(e^{j\omega})\Big|_{\omega = \frac{2\pi k}{N}} \quad (A.6)$$

$H[k]$ can be seen to be the samples of the Fourier transform of $h[n]$. The reciprocal of these values are samples of the Fourier transform of $g[n]$:

$$\frac{1}{H[k]} = G(e^{j\omega})\Big|_{\omega = \frac{2\pi k}{N}} \quad (A.7)$$

Because $g[n]$ has infinite time support, the discrete spectrum in equation A.7 is undersampled, and if we compute the inverse DFT, the result will be time aliased:

$$\tilde{g}[n] = \sum_{r=-\infty}^{\infty} g[n+rN] = IDFT\left(\frac{1}{H[k]}\right) \tag{A.8}$$

Because $g[n]$ decays exponentially with increasing $|n|$, as we increase the size of the DFT, the alias distortion will decrease, and $\tilde{g}[n]$ will approach $g[n]$. When $H(z)$ has zeros near the unit circle, this produces poles in $G(z)$ that decay slowly, requiring a larger DFT size to combat aliasing. A practical benefit of the DFT method is that we can limit the magnitude of $|1/H[k]|$ prior to the IDFT operation. This limits the gain of the inverse filter, and allows us to compute $\tilde{g}[n]$ using a smaller DFT size.

Summarizing, we can compute an approximate inverse filter for a finite length signal $h[n]$ following these steps:

1. Compute the DFT of $h[n]$ (equation A.6) using the fast Fourier transform (FFT), padding $h[n]$ with zeros as necessary.
2. Compute $1/H[k]$, and limit the resulting magnitudes.
3. Compute the IDFT of the magnitude limited spectrum using the inverse FFT.
4. Window the periodic aliased result (equation A.5) and delay to obtain a causal filter. Alternatives to a rectangular window (e.g., a Hanning window) may be used to smooth the transient behavior of the inverse filter.

This procedure yields an FIR filter that implements an approximate inverse cascaded with a modeling delay.

REFERENCES

Abe, K., F. Asano, Y. Suzuki, and T. Sone (1995). "Sound pressure control at multiple points for sound reproduction", *Proc. Int. Congress on Acoust.*, Trondheim, Norway.

AES (1986). *Stereophonic Techniques*, The Audio Engineering Society, New York, NY.

Asano, F., Y. Suzuki, and T. Sone (1990). "Role of spectral cues in median plane localization", *J. Acoust. Soc. Am.*, 88(1), pp. 159-168.

Asano, F., Y. Suzuki, and T. Sone (1996). "Sound equalization using derivative constraints", *Acustica*, 82, pp. 311-320.

Atal, B. S., and M. Schroeder (1966). Apparent sound source translator. United States Patent 3,236,949.

Bamford, J. S., and J. Vanderkooy (1995). "Ambisonic Sound for Us", *Proc. Audio Eng. Soc. Conv.* Preprint 4138.

Basu, S., I. Essa, and A. Pentland (1996). "Motion Regularization for Model-based Head Tracking", MIT Media Lab, Perceptual Computing Section, Technical Report no. 362.
<ftp://whitechapel.media.mit.edu/pub/tech-reports/TR-362.ps.Z>

Batteau, D. W. (1967). "The role of the pinna in human localization", *Proc. Royal Soc. London*, Ser. B, 168, pp. 158-180.

Bauck, J., and D. H. Cooper (1992). "Generalized Transaural Stereo", *Proc. Audio Eng. Soc. Conv.* Preprint 3401.

Bauck, J., and D. H. Cooper (1993). "Developments in Transaural Stereo", *Proc. IEEE ASSP Workshop on Applications of Signal Processing to Audio and Acoustics*, New Paltz, New York.

Bauck, J., and D. H. Cooper (1996). "Generalized Transaural Stereo and Applications", *J. Audio Eng. Soc.*, 44(9), pp. 683-705.

Bauer, B. B. (1961a). "Phasor Analysis of Some Stereophonic Phenomena", *J. Acoust. Soc. Am.*, 33(11), pp. 1536-1539.

Bauer, B. B. (1961b). "Stereophonic Earphones and Binaural Loudspeakers", *J. Audio Eng. Soc.*, 9(2), pp. 148-151.

Begault, D. R. (1990). "Challenges to the Successful Implementation of 3-D Sound", *J. Audio Eng. Soc.*, 39(11), pp. 864-870.

Begault, D. R. (1992). "Perceptual effects of synthetic reverberation on three-dimensional audio systems", *J. Audio Eng. Soc.*, 40, pp. 895-904.

Begault, D. R. (1994). *3-D Sound for Virtual Reality and Multimedia*, Academic Press, Cambridge, MA.

Berkhout, A. J. (1988). "A Holographic Approach to Acoustic Control", *J. Audio Eng. Soc.*, 36(12), pp. 977-995.

Berkhout, A. J., D. de Vries, and P. Vogel (1993). "Acoustic control by wave field synthesis", *J. Acoust. Soc. Am.*, 93, pp. 2764-2778.

Blauert, J. (1969/70). "Sound Localization in the Median Plane", *Acustica*, 22, pp. 205-213.

Blauert, J. (1983). *Spatial Hearing*, MIT Press, Cambridge, MA.

Bloom, P. J. (1977). "Determination of monaural sensitivity changes due to the pinna by use of minimum-audible-field measurements in the lateral vertical plane", *J. Acoust. Soc. Am.*, 61(3), pp. 820-828.

Blumlein, A. D. (1933). British Patent 394,325. Reprinted in (Blumlein, 1958) and (AES, 1986).

Blumlein, A. D. (1958). "British Patent 394,325", *J. Audio Eng. Soc.*, 6, pp. 91-98. Reprint of (Blumlein, 1933), also in (AES, 1986).

Boerger, G., P. Laws, and J. Blauert (1977). "Stererophonic Reproduction by Earphones with Control of Special Transfer Functions though Head Movements", *Acustica*, 39, pp. 22-26.

Boone, M. M., E. N. G. Verheijen, and P. F. V. Tol (1995). "Spatial Sound-Field Reproduction by Wave Field Synthesis", *J. Audio Eng. Soc.*, 43(12), pp. 1003-1012.

Bregman, A. S. (1990). *Auditory scene analysis*, MIT Press, Cambridge, MA.

Burkhard, M. D., and R. M. Sachs (1975). "Anthropometric manikin for acoustic research", *J. Acoust. Soc. Am.*, 58, pp. 214-222.

Burraston, D. M., M. P. Hollier, and M. O. Hawksford (1997). "Limitations of Dynamically Controlling the Listening Position In a 3-D Ambisonic Environment", *Proc. Audio Eng. Soc. Conv.* Preprint 4460.

Burrus, C. S., and T. W. Parks (1970). "Time Domain Design of Recursive Digital Filters", *IEEE Trans. Audio and Electroacoust.*, AU-18(2), pp. 137-141.

Burrus, C. S., and T. W. Parks (1987). *Digital Filter Design*, Wiley, New York.

Butler, R. A., and K. Belendiuk (1977). "Spectral cues utilized in the localization of sound in the median saggital plane", *J. Acoust. Soc. Am.*, 61(5), pp. 1264-1269.

Butler, R. A. (1997). "Spatial Referents of Stimulus Frequencies: Their Role in Sound Localization", in *Binaural and Spatial Hearing in Real and Virtual Environments*, Gilkey, R. H., and T. R. Anderson, Eds., Lawrence Erlbaum Associates, Mahwah, NJ.

Casey, M. A., W. G. Gardner, and S. Basu (1995). "Vision Steered Beam-Forming and Transaural Rendering for the Artificial Life Interactive Video Environment (ALIVE)", *Proc. Audio Eng. Soc. Conv.* Preprint 4052.

Cohen, J. M. (1982). Stereo image separation and perimeter enhancement. U.S. Patent 4,355,203.

Cooper, D. H. (1982). "Calculator Program for Head-Related Transfer Function", *J. Audio Eng. Soc.*, 30(1/2), pp. 34-38.

Cooper, D. H. (1987). "Problems with Shadowless Stereo Theory: Asymptotic Spectral Status", *J. Audio Eng. Soc.*, 35(9), pp. 629-642.

Cooper, D. H., and J. L. Bauck (1989). "Prospects for Transaural Recording", *J. Audio Eng. Soc.*, 37(1/2), pp. 3-19.

Cooper, D. H., and J. L. Bauck (1990). Head diffraction compensated stereo system. U.S. Patent 4,893,342.

Craven, P. G., and M. A. Gerzon (1992). "Practical Adaptive Room and Loudspeaker Equaliser for Hi-Fi Use", *Proc. Audio Eng. Soc. Conv.* Preprint 3346.

Damaske, P., and V. Mellert (1969). "A procedure for generating directionally accurate sound images in the upper half-space using two loudspeakers", *Acustica*, 22, pp. 154-162.

Damaske, P. (1971). "Head-related Two-channel Stereophony with Loudspeaker Reproduction", *J. Acoust. Soc. Am.*, 50(4), pp. 1109-1115.

Duda, R. O. (1997). "Elevation Dependence of the Interaural Transfer Function", in *Binaural and Spatial Hearing in Real and Virtual Environments*, Gilkey, R. H., and T. R. Anderson, Eds., Lawrence Erlbaum Associates, Mahwah, NJ.

Durlach, N. I. (1968). A Decision Model for Psychophysics. Unpublished.

Durlach, N. I., and S. Colburn (1978). "Binaural Phenomena", in *Handbook of Perception, Vol. IV*, Carterette, E., Ed., Academic Press.

Durlach, N. I., A. Rigopulos, X. D. Pang, W. S. Woods, A. Kulkarni, H. S. Colburn, and E. M. Wenzel (1992). "On the Externalization of Auditory Images", *Presence*, 1(2), pp. 251-257.

Elliot, S. J., and P. A. Nelson (1985). "Algorithm for Multichannel LMS Adaptive Filtering", *Electronics Letters*, 21(21), pp. 979-981.

Elliot, S. J., I. M. Strothers, and P. A. Nelson (1987). "A Multiple Error LMS Algorithm and Its Application to the Active Control of Sound and Vibration", *IEEE Trans. Acoust,. Speech, and Signal Processing*, 35(10), pp. 1423-1434.

Elliot, S. J., and P. A. Nelson (1989). "Multiple-Point Equalization in a Room Using Adaptive Digital Filters", *J. Audio Eng. Soc.*, 37(11), pp. 899-907.

Farrar, K. (1979). "Soundfield Microphone", *Wireless World*, 85, pp. 99-102.

Fisher, N. I., T. Lewis, and B. J. J. Embleton (1987). *Statistical Analysis of Spherical Data*, Cambridge Univ. Press, Cambridge, United Kingdom.

Friedlander, B., and B. Porat (1984). "The Modified Yule-Walker Method of ARMA Spectral Estimation", *IEEE Trans. Aerospace and Electronics Systems*, 20(2), pp. 158-173.

Gardner, M. B., and R. S. Gardner (1973). "Problem of localization in the median plane: effect of pinnae cavity occlusion", *J. Acoust. Soc. Am.*, 53(2), pp. 400-408.

Gardner, M. B. (1973). "Some monaural and binaural facets of median plane localization", *J. Acoust. Soc. Am.*, 54(6), pp. 1489-1495.

Gardner, W. G. (1992). *The Virtual Acoustic Room*, Master's thesis, MIT Media Lab.

Gardner, W. G. (1995). "Transaural 3-D audio", M.I.T. Media Lab, Perceptual Computing Section, Technichal Report no. 342.
<http://sound.media.mit.edu/papers.html>

Gardner, W. G., and K. D. Martin (1994). "HRTF measurements of a KEMAR Dummy-Head Microphone", M.I.T. Media Lab, Perceptual Computing Section, Technical Report no. 280.
<http://sound.media.mit.edu/KEMAR.html>

Gardner, W. G., and K. D. Martin (1995). "HRTF measurements of a KEMAR", *J. Acoust. Soc. Am.*, 97(6), pp. 3907-3908.

Gardner, W. G. (1997). "Head Tracked 3-D Audio Using Loudspeakers", *Proc. IEEE Workshop on Applications of Signal Processing to Audio and Acoustics*, New Paltz, NY.

Gardner, W. G. (1998). "Reverberation Algorithms", in *Applications of Signal Processing to Audio and Acoustics*, Kahrs, M., and K. Brandenburg, Eds., Kluwer Academic Publishers, Norwell, MA.

Gerzon, M. A. (1985). "Ambisonics in Multichannel Broadcasting and Video", *J. Audio Eng. Soc.*, 33(11), pp. 859-871.

Gerzon, M. A. (1992). "Psychoacoustic Decoders for Multispeaker Stereo and Surround Sound", *Proc. Audio Eng. Soc. Conv.* Preprint 3406.

Gerzon, M. A. (1994). "Applications of Blumlein Shuffling to Stereo Microphone Techniques", *J. Audio Eng. Soc.*, 42(6), pp. 435-453.

Gilkey, R. H., and T. R. Anderson, Eds. (1997). *Binaural and Spatial Hearing in Real and Virtual Environments*, Lawrence Erlbaum Associates, Mahwah, NJ.

Griesinger, D. (1986). "Spaciousness and Localization in Listening Rooms and Their Effects on the Recording Technique", *J. Audio Eng. Soc.*, 34(4), pp. 255-268.

Griesinger, D. (1987). "New perspectives on coincident and semi-coincident microphone arrays", *Proc. Audio Eng. Soc. Conv.* Preprint 1987.

Griesinger, D. (1989a). "Equalization and Spatial Equalization of Dummy-Head Recordings for Loudspeaker Reproduction", *J. Audio Eng. Soc.*, 37(1/2), pp. 20-29.

Griesinger, D. (1989b). "Theory and Design of a Digital Audio Signal Processor for Home Use", *J. Audio Eng. Soc.*, 37(1/2), pp. 40-50.

Griesinger, D. (1991). "Improving Room Acoustics Through Time-Variant Synthetic Reverberation", *Proc. Audio Eng. Soc. Conv.* Preprint 3014.

Griesinger, D. (1996). "Multichannel Matrix Surround Decoders for Two-Eared Listeners", *Proc. Audio Eng. Soc. Conv.* Preprint 4402.

Hartmann, W. M. (1983). "Localization of sound in rooms", *J. Acoust. Soc. Am.*, 74(5), pp. 1380-1391.

Hartmann, W. M., and A. Wittenberg (1996). "On the externalization of sound images", *J. Acoust. Soc. Am.*, 99(6), pp. 3678-3688.

Hartmann, W. M. (1997). "Listening in a Room and the Precedence Effect", in *Binaural and Spatial Hearing in Real and Virtual Environments*, Gilkey, R. H., and T. R. Anderson, Eds., Lawrence Erlbaum Associates, Mahwah, NJ.

Hebrank, J., and D. Wright (1974a). "Are two ears necessary for localization of sound sources on the median plane?", *J. Acoust. Soc. Am.*, 56(3), pp. 935-938.

Hebrank, J., and D. Wright (1974b). "Spectral cues used in the localization of sound sources on the median plane", *J. Acoust. Soc. Am.*, 56(6), pp. 1829-1834.

Heegaard, F. D. (1992). "The Reproduction of Sound in Auditory Perspective and a Compatible System of Stereophony", *J. Audio Eng. Soc.*, 40(10), pp. 802-808.

Horn, R. A., and C. R. Johnson (1985). *Matrix Analysis*, Cambridge Univ. Press, Cambridge, United Kingdom.

Howell, D. C. (1997). *Statistical Methods for Psychology*, Duxbury Press.

Iwahara, M., and T. Mori (1978). Stereophonic sound reproduction system. United States Patent 4,118,599.

Jot, J.-M. (1992). *Etude et réalisation d'un spatialisateur de sons par modèles physiques et perceptifs*, Ph.D. thesis, Telecom Paris.

Jot, J.-M., V. Larcher, and O. Warusfel (1995). "Digital signal processing issues in the context of binaural and transaural stereophony", *Proc. Audio Eng. Soc. Conv.* Preprint 3980. See (Jot et al., 1997) for revised edition.

Jot, J.-M. (1996). "Synthesizing three-dimensional sound scenes in audio or multimedia production and interactive human-computer interfaces", Presented at 5th Int. Conf., Interface to real and virtual worlds, Montpellier, France.

Jot, J.-M. (1997). "Real-time spatial processing of sounds for music, multimedia, and interactive human-computer interfaces", submitted to *ACM Multimedia Systems Journal* (special issue on audio and multimedia).

Jot, J.-M., V. Larcher, and O. Warusfel (1997). "Digital signal processing issues in the context of head-related stereophony", to appear in *J. Audio Eng. Soc.*

Kistler, D. J., and F. L. Wightman (1992). "A model of head-related transfer function based on principal component analysis and minimum-phase reconstruction", *J. Acoust. Soc. Am.*, 91(3), pp. 1637-1647.

Kleiner, M. (1981). "Speech Intelligibility in Real and Simulated Sound Fields", *Acustica*, 47(2).

Koring, J., and A. Schmitz (1993). "Simplifying Cancellation of Cross-Talk for Playback of Head-Related Recordings in a Two-Speaker System", *Acustica*, 79, pp. 221-232.

Kotorynski, K. (1990). "Digital Binaural/Stereo Conversion and Crosstalk Cancelling", *Proc. Audio Eng. Soc. Conv.* Preprint 2949.

Kuhn, G. F. (1977). "Model for the interaural time differences in the azimuthal plane", *J. Acoust. Soc. Am.*, 62, pp. 157-167.

Kuhn, G. F. (1987). "Physical Acoustics and Measurements Pertaining to Directional Hearing", in *Directional Hearing*, Yost, W. A., and G. Gourevitch, Eds., Springer-Verlag, New York, NY.

Laakso, T. I., V. Valimaki, M. Karjalainen, and U. K. Laine (1996). "Splitting the Unit Delay", *IEEE Signal Proc. Mag.*, 13(1), pp. 30-60.

Larcher, V., and J.-M. Jot (1997a). "Techniques d'interpolation de filtres audio-numeriques. Application a la reproduction spatiale des sons sur ecouteurs", *Proc. of the 4th congress of the French Soc. of Acoust*

Larcher, V., and J.-M. Jot (1997b). HRIR measurements of a Head Acoustics HMSII manikin and five human subjects. Unpublished.

Lipshitz, S. P. (1986). "Stereo Microphone Techniques... Are the Purists Wrong?", *J. Audio Eng. Soc.*, 34(9), pp. 716-744.

Lopez-Poveda, E. A. (1996). *The physical origin and physiological coding of pinna-based spectral cues*, Ph.D. dissertation, Loughborough Univ.

Lopez-Poveda, E. A., and R. Meddis (1996). "A physical model of sound diffraction and reflections in the human concha", *J. Acoust. Soc. Am.*, 100(5), pp. 3248-3259.

Makous, J. C., and J. C. Middlebrooks (1990). "Two-dimensional sound localization by human listeners", *J. Acoust. Soc. Am.*, 87(5), pp. 2188-2200.

Mehrgardt, S., and V. Mellert (1977). "Transformation characteristics of the external human ear", *J. Acoust. Soc. Am.*, 61(6), pp. 1567-1576.

Meyer, von E., W. Burgtorf, and P. Damaske (1965). "Eine Apparatur Zur Elektroakustischen Nachbildung Von Schallfeldern", *Acustica*, 15.

Meyer, K., H. L. Applewhite, and F. A. Biocca (1993). "A survey of position trackers", *Presence*, 1, pp. 173-200.

Middlebrooks, J. C., and D. M. Green (1991). "Sound localization by human listeners", *Annu. Rev. Psychol.*, 42, pp. 135-159.

Middlebrooks, J. C. (1992). "Narrow-band sound localization related to external ear acoustics", *J. Acoust. Soc. Am.*, 92(5), pp. 2607-2624.

Middlebrooks, J. C. (1997). "Spectral Shape Cues for Sound Localization", in *Binaural and Spatial Hearing in Real and Virtual Environments*, Gilkey, R. H., and T. R. Anderson, Eds., Lawrence Erlbaum Associates, Mahwah, NJ.

Mills, W. A. (1972). "Auditory Localization", in *Foundations of Modern Auditory Theory*, vol. II, Tobias, J. V., Ed., Academic Press, New York.

Miyoshi, M., and Y. Kaneda (1988). "Inverse Filtering of Room Acoustics", *IEEE Trans. Acoust,. Speech, and Signal Processing*, 36(2), pp. 145-152.

Møller, H. (1989). "Reproduction of Artificial-Head Recordings through Loudspeakers", *J. Audio Eng. Soc.*, 37(1/2), pp. 30-33.

Møller, H. (1992). "Fundamentals of Binaural Technology", *Applied Acoustics*, 36, pp. 171-218.

Møller, H., D. Hammershøi, C. B. Jensen, and M. F. Sørensen (1995a). "Transfer Characteristics of Headphones Measured on Human Ears", *J. Audio Eng. Soc.*, 43(4), pp. 203-217.

Møller, H., C. B. Jensen, D. Hammershøi, and M. F. Sørensen (1995b). "Design Criteria for Headphones", *J. Audio Eng. Soc.*, 43(4), pp. 218-232.

Møller, H., M. F. Sørensen, D. Hammershøi, and C. B. Jensen (1995c). "Head-Related Transfer Functions of Human Subjects", *J. Audio Eng. Soc.*, 43(5), pp. 300-321.

Møller, H., M. F. Sørensen, C. B. Jensen, and D. Hammershøi (1996a). "Binaural Technique: Do We Need Individual Recordings?", *J. Audio Eng. Soc.*, 44(6), pp. 451-469.

Møller, H., C. B. Jensen, D. Hammershøi, and M. F. Sørensen (1996b). "Using a Typical Human Subject for Binaural Recording", *Proc. Audio Eng. Soc. Conv.* Preprint 4157.

Morimoto, M., and Z. Maekawa (1988). "Effects of Low Frequency Components on Auditory Spaciousness", *Acustica*, 66, pp. 190-196.

Morse, P. M., and K. U. Ingard (1968). *Theoretical Acoustics*, McGraw-Hill, New York, NY.

Mourjopoulos, J. (1985). "On the variation and invertibility of room impulse response functions", *J. Sound and Vibration*, 102(2), pp. 217-228.

Neely, S. T., and J. B. Allen (1979). "Invertibility of a room impulse response", *J. Acoust. Soc. Am.*, 66(1), pp. 165-169.

Nelson, P. A., H. Hamada, and S. J. Elliot (1992). "Adaptive Inverse Filters for Stereophonic Sound Reproduction", *IEEE Trans. Signal Processing*, 40(7), pp. 1621-1632.

Nelson, P. A., F. Orduna-Bustamante, and H. Mamada (1995). "Inverse Filter Design and Equalization Zones in Multichannel Sound Reproduction", *IEEE Trans. Speech and Audio Processing*, 3(3), pp. 185-192.

Oldfield, S. R., and S. P. A. Parker (1984a). "Acuity of sound localisation: a topography of auditory space. I. Normal hearing conditions", *Perception*, 13, pp. 581-600.

Oldfield, S. R., and S. P. A. Parker (1984b). "Acuity of sound localisation: a topography of auditory space. II. Pinna cues absent", *Perception*, 13, pp. 601-617.

Oldfield, S. R., and S. P. A. Parker (1986). "Acuity of sound localisation: a topography of auditory space. III. Monaural hearing conditions", *Perception*, 15, pp. 67-81.

Oliver, N., A. P. Pentland, and F. Berard (1997). "LAFTER: Lips and Face Real Time Tracker", *Proc. IEEE Int. Conf. on Computer Vision and Pattern Recognition*.

Olson, H. F. (1972). *Modern Sound Reproduction*, Van Nostrand Reinhold, New York, NY.

Oppenheim, A. V., and R. W. Schafer (1989). *Discrete Time Signal Processing*, Prentice Hall, Englewood Cliffs, NJ.

Papoulis, A. (1991). *Probability, Random Variables, and Stochastic Processes*, McGraw-Hill, New York, NY.

Plenge, G. (1974). "On the differences between localization and lateralization", *J. Acoust. Soc. Am.*, 56(3), pp. 944-951.

Pulkki, V. (1997). "Virtual Sound Source Positioning Using Vector Base Amplitude Panning", *J. Audio Eng. Soc.*, 45(6), pp. 456-466.

Rakerd, B., and W. M. Hartmann (1983). "Localization of sound in rooms, II: The effects of a single reflecting surface", *J. Acoust. Soc. Am.*, 78(2), pp. 524-533.

Lord Rayleigh [Strutt, J. W.] (1907). "On our Perception of Sound Direction", *Phil. Mag.*, 13, pp. 214-232.

Lord Rayleigh [Strutt, J. W.] (1945). *The Theory of Sound* (2nd Ed), Dover Publications, New York. Originally published in 1877.

Sakamoto, N., T. Gotoh, T. Kogure, M. Shimbo, and A. H. Clegg (1981). "Controlling Sound-Image Localization in Stereophonic Reproduction", *J. Audio Eng. Soc.*, 29(11), pp. 794-799.

Sakamoto, N., T. Gotoh, T. Kogure, M. Shimbo, and A. H. Clegg (1982). "Controlling Sound-Image Localization in Stereophonic Reproduction: Part II", *J. Audio Eng. Soc.*, 30(10), pp. 719-722.

Schroeder, M. R., and B. S. Atal (1963). "Computer simulation of sound transmission in rooms", *IEEE Int. Conv. Record*, 7, pp. 150-155.

Schroeder, M. R. (1970). "Digital simulation of sound transmission in reverberant spaces", *J. Acoust. Soc. Am.*, 47(2), pp. 424-431.

Schroeder, M. R. (1973). "Computer Models for Concert Hall Acoustics", *Am. J. Physics*, 41, pp. 461-471.

Schroeder, M. R., D. Gottlob, and K. F. Siebrasse (1974). "Comparative Stude of European Concert Halls", *J. Acoust. Soc. Am.*, 56, pp. 1195-1201.

Searle, C. L., L. D. Braida, D. R. Cuddy, and M. F. Davis (1975). "Binaural pinna disparity: another auditory localization cue", *J. Acoust. Soc. Am.*, 57(2), pp. 448-455.

Shaw, E. A. G. (1966). "Earcanal Pressure Generated by a Free Sound Field", *J. Acoust. Soc. Am.*, 39(3), pp. 465-470.

Shaw, E. A. G. (1974). "Transformation of sound pressure from the free field to the eardrum in the horizontal plane", *J. Acoust. Soc. Am.*, 56(6), pp. 1848-1861.

Shaw, E. A. G., and R. Teranishi (1968). "Sound Pressure Generated in an External-Ear Replica and Real Human Ears by a Nearby Point Source", *J. Acoust. Soc. Am.*, 44(1), pp. 240-249.

Shaw, E. A. G., and M. M. Vaillancourt (1985). "Transformation of sound-pressure level from the free field to the eardrum presented in numerical form", *J. Acoust. Soc. Am.*, 78(3), pp. 1120-1123.

Shinn-Cunningham, B., H. Lehnert, G. Kramer, E. Wenzel, and N. Durlach (1997). "Auditory Displays", in *Binaural and Spatial Hearing in Real and Virtual Environments*, Gilkey, R. H., and T. R. Anderson, Eds., Lawrence Erlbaum Associates, Mahwah, NJ..

Smith, J. O., and J. S. Abel (1995). "The Bark Bilinear Transform", *Proc. IEEE Workshop on Applications of Signal Processing to Audio and Acoustics*, New Paltz, NY.

Start, E. W., V. G. Valstar, and D. de Vries (1995). "Application of Spacial Bandwidth Reduction in Wave Field Synthesis", *Proc. Audio Eng. Soc. Conv.* Preprint 3972.

Theile, G., and G. Plenge (1977). "Localization of Lateral Phantom Sources", *J. Audio Eng. Soc.*, 25(4).

Theile, G. (1986). "On the Standardization of the Frequency Response of High-Quality Studio Headphones", *J. Audio Eng. Soc.*, 34, pp. 956-969.

Theile, G. (1993). "The New Sound Format "3/2-Stereo"", *Proc. Audio Eng. Soc. Conv.*

Thurlow, W. R., and P. S. Runge (1967). "Effect of Induced Head Movements on Localization of Direction of Sounds", *J. Acoust. Soc. Am.*, 42(2), pp. 480-488.

Thurlow, W. R., J. W. Mangels, and P. S. Runge (1967). "Head Movements During Sound Localization", *J. Acoust. Soc. Am.*, 42(2), pp. 489-493.

Toole, F. E. (1970). "In-Head Localization of Acoustic Images", *J. Acoust. Soc. Am.*, 4(2), pp. 943-949.

Torkkola, K. (1996). "Blind separation of convolved sources based on information maximization", *Proc. IEEE Workshop on Neural Networks for Signal Processing*, Kyoto, Japan.

Wallach, H. (1939). "On Sound Localization", *J. Acoust. Soc. Am.*, 10, pp. 270-274.

Wallach, H. (1940). "The Role of Head Movements and Vestibular and Visual Cues in Sound Localization", *J. Exptl. Psychol.*, 27, pp. 339-346.

Weiss, L., and R. N. McDonough (1963). "Prony's method, z-transforms, and Pade approximation", *SIAM Rev.*, 3(2), pp. 145-149.

Wenzel, E. M. (1992). "Localization in Virtual Acoustic Displays", *Presence*, 1(1), pp. 80-107.

Wenzel, E. M., M. Arruda, D. J. Kistler, and F. L. Wightman (1993). "Localization using nonindividualized head-related transfer functions", *J. Acoust. Soc. Am.*, 94(1), pp. 111-123.

Wenzel, E. M. (1995). "The relative contribution of interaural time and magnitude cues to dynamic sound localization", *Proc. IEEE Workshop on Applications of Signal Processing to Audio and Acoustics*, New Paltz, NY.

Wightman, F. L., and D. J. Kistler (1989a). "Headphone simulation of free-field listening. I: Stimulus synthesis", *J. Acoust. Soc. Am.*, 85(2), pp. 858-867.

Wightman, F. L., and D. J. Kistler (1989b). "Headphone simulation of free-field listening. II: Psychophysical validation", *J. Acoust. Soc. Am.*, 85(2), pp. 868-878.

Wightman, F., and D. Kistler (1992). "The dominant role of low-frequency interaural time differences in sound localization", *J. Acoust. Soc. Am.*, 91(3), pp. 1648-1661.

Wightman, F., and D. Kistler (1997). "Salience of Sound Localization Cues", in *Binaural and Spatial Hearing in Real and Virtual Environments*, Gilkey, R. H., and T. R. Anderson, Eds., Lawrence Erlbaum Associates, Mahwah, NJ.

Yanagida, M., O. Kakusho, and T. Gotoh (1983). "Application of the least-squares method to sound-image localization in multi-loudspeaker multi-listener case", *J. Acoust. Soc. Jpn.*, 4(2), pp. 107-109.

Yost, W. A., and G. Gourevitch, Ed (1987). *Directional Hearing*, Springer-Verlag, New York, NY.

Zurek, P. M. (1987). "The Precedence Effect", in *Directional Hearing*, Yost, W. A., and G. Gourevitch, Eds., Springer-Verlag, New York, NY.

INDEX

A

Acoustical measurement 89–97
Acoustical simulation 80–87
Acoustical transfer matrix 47
Allpass system 39
Ambisonics 17
Auditory fusion 127
Average angle error 105
Azimuth localization 106, 117, 121, 124, 129

B

Bark bilinear transform 43
Binaural audio 2
Binaural signal 45
Binaural synthesis 2, 21, 45, 77
Boosted band 9, 37, 114

C

Causality 16, 40, 141
Channel separation 80–87
Commutation 64
Concha 9, 29
Cone of confusion 8, 54
Cross-correlation 31
Crosstalk cancellation 2, 13–16, 44–78
Crosstalk canceller 13, 21, 46, 77
 asymmetric 4, 5, 15, 60–64, 67
 bandlimited 3, 59–70
 individualized 15, 131, 137
 lattice 13, 15, 49, 67
 non-individualized 3, 16, 115
 parameter dependency on head position 64–67
 recursive 4, 14, 54–56, 60–64
 shuffler 14, 15, 57, 68
 symmetric 4, 5, 56–59, 68–70, 132

D

Delay line 64
Determinant 49

Diffuse-field average 34
Diffuse-field equalization 28, 34, 101
Distance 8, 109
Distant listening 6, 136
Duplex theory 7

E

Ear canal resonance 29
Elevation localization 8, 9, 10, 111
Equalization zone 3, 4, 5, 16, 17, 83–87, 116, 120
Excess phase 39
Externalization 2, 10, 109

F

Finite impulse response (FIR) filter 17, 68, 123, 127
 inverse of 141–143
Free-field equalization 27, 35, 58, 62

H

Head diameter 31, 93
Head diffraction 2, 8
 also see Interaural transfer function
Head model 4, 21, 40, 70, 81, 133, 137
 spherical 14, 31, 66, 93
Head motion 11–12
Head related impulse response (HRIR) 24
Head related transfer function (HRTF) 2
 equalization 26–29, 34
 individualized 10, 15
 KEMAR 24–44
 measurement 24–26
 non-individualized 3, 10, 11
 structural averaging 138
Head rotation 3, 11, 123–131
Head tracking 3, 4, 5, 11, 12, 22, 77, 116, 128, 140
Head transfer matrix 47

Head translation 3, 87, 116–123
Headphones 2, 10, 101
Highpass filter 59, 76
Horizontal diffuse-field equalization 28

I

Infinite impulse response (IIR) filter 42, 69
Intensity panning 12, 19
Interaural difference spectra 9
Interaural level difference (ILD) 7, 12, 30, 75
Interaural phase delay 38
Interaural time delay (ITD) 7, 12, 31, 40, 93
Interaural transfer function (ITF) 37–44, 50
Interpolation 44, 64
Inverse filtering of room acoustics 16–17, 139
Inverse kappa 105
Ipsilateral equalization 50, 62

K

KEMAR 3, 24, 89

L

Lagrangian interpolation 66
Lateral (LR) angle 106, 107, 117, 122, 125, 130
Loudspeaker 2, 11, 15, 104
 displays 12–19
 placement 138
Loudspeaker binaural audio 3
Loudspeaker binaural signal 46
Lowpass filter 37, 40, 59, 70, 76

M

Maximum length (ML) sequence 25
Measurement equalization 27
Microphone 4, 17, 24, 28, 91
Minimum-phase 15, 16, 35, 39, 141
Modeling delay 16, 40, 48, 62, 63, 142
Multichannel system 16, 17, 17–19, 23, 139
Multiple listeners 2, 5, 16, 135

P

Phasor analysis 12, 88
Pinna 2, 9, 10, 11, 24, 89, 91
Power compensation model 3, 70–76, 132, 139
Power panning property 23, 71
Precedence mechanism 136
Prony's method 43, 69

Q

Quadraphonic sound 18

R

Realizability 51–54

Reverberation 1, 10, 11, 15, 19, 22, 78
Reversal 2, 8, 10, 12, 14, 15, 16, 105, 106, 129

S

Shelving filter 17, 74, 76
Shuffler filter 14, 15, 57
 also see Crosstalk canceller
Sound localization blur 105
Sound localization by humans 7–12
Sound localization cue
 dynamic 3, 10–12, 132
 interaural 7–8, 11
 spectral 9–10
Sound localization experiments 99–132
 baseline results 106–115
 charts used 101
 dynamic head motion 127–131
 error statistics 108, 119, 122, 125, 129
 front-back head translation 120–123
 head rotation 123–127
 headphone procedure 100–103
 lateral head translation 116–120
 loudspeaker procedure 103–105
 sighting aid 104
 statistical analysis 105–106
Spatial auditory display 1–3, 4, 7–19
Speaker and air transfer matrix 47
Spectral notch 31, 138
Stability 16, 51–54, 141
Stereo 1, 12–13
Stimulus 8, 9, 11, 100, 103, 123, 128
Summing localization 13
Surround sound system 2, 18

T

T test 118
Time panning 12
Transmission path inversion 2, 16, 48

W

Wave field synthesis 19
Wavelet 27
Window 27, 92, 142

Y

Yule-Walker method 43